ARACHIDONIC ACID METABOLISM AND TUMOR PROMOTION

PROSTAGLANDINS, LEUKOTRIENES, AND CANCER

Series Editors: Kenneth V. Honn and Lawrence J. Marnett
Wayne State University School of Medicine
Detroit, Michigan

W.E.M. Lands, ed.: Biochemistry of Arachidonic Acid Metabolism. 1985. ISBN 0-89838-717-5.
L.J. Marnett, ed.: Arachidonic Acid Metabolism and Tumor Initiation. 1985. ISBN 0-89838-729-9.
S.M. Fischer and T.J. Slaga, eds.: Arachidonic Acid Metabolism and Tumor Promotion. 1985. ISBN 0-89838-724-8.
J.S. Goodwin, ed.: Prostaglandins and Immunity. 1985. ISBN 0-89838-723-X.

ARACHIDONIC ACID METABOLISM AND TUMOR PROMOTION

edited by

Susan M. Fischer
and
Thomas J. Slaga
The University of Texas System Cancer Center
Smithville, Texas

Martinus Nijhoff Publishing
a member of the Kluwer Academic Publishers Group
Boston/Dordrecht/Lancaster

Distributors for North America:
Kluwer Academic Publishers
190 Old Derby Street
Hingham, MA 02043

Distributors for all other countries:
Kluwer Academic Publishers Group
Distribution Centre
P.O. Box 322
3300 AH Dordrecht
The Netherlands

Library of Congress Cataloging in Publication Data

Main entry under title:

Arachidonic acid metabolism and tumor promotion.

(Prostaglandins, leukotrienes, and cancer)
Includes bibliographies and index.
1. Cocarcinogenesis. 2. Arachidonic acid—
Metabolism. I. Fischer, Susan M. II. Slaga,
Thomas J. III. Series. [DNLM: 1. Arachidonic
Acids—metabolism. 2. Neoplasms—etiology.
QU 90 A658]
RC268.52.A73   1985   616.99'4071   85-4968
ISBN 0-89838-724-8

Copyright © 1985 by Martinus Nijhoff Publishing, Boston

All rights reserved. No part of this publication may be reproduced, stored in a retrieval system, or transmitted in any form or by any means, mechanical, photocopying, recording, or otherwise, without written permission of the publisher, Martinus Nijhoff Publishing, 190 Old Derby Street, Hingham, Massachusetts 02043.

Printed in the United States of America

CONTENTS

CONTRIBUTORS		vi
FOREWORD		vii
PREFACE		viii
INTRODUCTION		1
1.	ARACHIDONIC ACID METABOLISM IN THE SKIN Vincent A. Ziboh	4
2.	ARACHIDONIC ACID METABOLISM AND TUMOR PROMOTION Susan M. Fischer	21
3.	PROSTAGLANDINS, EPIDERMAL HYPERPLASIA AND SKIN TUMOR PROMOTION Gerhard Furstenberger and Friedrich Marks	49
4.	EFFECTS OF PHORBOL ESTERS ON PHOSPHOLIPID METABOLISM *IN VITRO* Philip W. Wertz	73
5.	TUMOR PROMOTING PHORBOL ESTERS MAY AFFECT CELL MEMBRANE SIGNAL TRANSMISSION BY PROTEIN KINASE Curtis L. Ashendel	101
6.	ACTIVE OXYGEN AND PROMOTION Peter A. Cerutti	131
7.	POSSIBLE INVOLVEMENT OF ARACHIDONATE PRODUCTS IN TUMOR PROMOTER INHIBITION OF CELL-CELL COMMUNICATION James E. Trosko, C. Aylsworth, C. Jone and C.C. Chang	169
8.	PROTEASES AND CYCLIC NUCLEOTIDES Sidney Belman and Seymour J. Garte	199
PERSPECTIVES Thomas J. Slaga and Susan M. Fischer		255
INDEX		259

CONTRIBUTORS

CURTIS L. ASHENDEL
Department of Medicinal Chemistry
Purdue University
Pharmacy Building
West Lafayette, Indiana 47907

CHARLES F. AYLSWORTH
Department of Anatomy
Michigan State University
East Lansing, Michigan 48824

SIDNEY BELMAN
Institute of Environmental Medicine
New York University Medical Center
550 First Avenue
New York, New York 10016

PETER A. CERUTTI
Department of Carcinogenesis
Swiss Institute for Experimental
  Cancer Research
Ch. de Boveresses
CH-1066 Epalinges/Lausanne
SWITZERLAND

CHIA CHENG CHANG
Department of Pediatrics and
  Human Development
B240 Life Science Building
Michigan State University
East Lansing, Michigan 48824

SUSAN M. FISCHER
The University of Texas
System Cancer Center
Science Park-Research Division
Post Office Box 389
Smithville, Texas 78957

GERHARD FURSTENBERGER
German Cancer Research Institute
Institute of Biochemistry
D-6900 Heidelberg
WEST GERMANY

SEYMOUR J. GARTE
New York University Medical Center
Department of Environmental Medicine
550 First Avenue
New York, New York 10016

CYRENIUS JONE
Department of Pediatrics and Human
  Development
B240 Life Science Building
Michigan State University
East Lansing, Michigan 48824

FRIEDRICH MARKS
German Cancer Research Institute
Institute of Biochemistry
D-6900 Heidelberg
WEST GERMANY

THOMAS J. SLAGA
The University of Texas
System Cancer Center
Science Park-Research Division
Post Office Box 389
Smithville, Texas 78957

JAMES E. TROSKO
Department of Pediatrics and
  Human Development
B240 Life Science Building
Michigan State University
East Lansing, Michigan 48824

PHILIP W. WERTZ
University of Iowa
Iowa City, Iowa 52242

VINCENT A. ZIBOH
Department of Dermatology
TB-192
University of California School
  of Medicine
Davis, California 95616

FOREWORD

Prostaglandins, Leukotrienes, and Cancer is a multi-volume series that will focus on an emerging area of cancer research. In 1968, R.H. Williams first reported that elevated prostaglandin levels are present in human medullary carcinoma. Since that time, the concept that arachidonic acid metabolites may be involved in cancer has expanded to include every aspect of the disease from cell transformation through metastasis.

Prostaglandins and leukotrienes are generic terms used to describe a family of bioactive lipids produced from unsaturated fatty acids (principally from arachidonic acid) via the cyclooxygenase and lipoxygenase pathways, respectively. Cyclooxygenase products consist of diverse products such as prostaglandin E_2 (PGE_2), prostacyclin (PGI_2) and thromboxane A_2 (TXA_2), whereas lipoxygenase products consist of hydroperoxy fatty acids and mono-, di- and tri-hydroxy acids including leukotrienes. The precursor fatty acids for the cyclooxygenase and lipoxygenase pathways are present in cellular phospholipids. This finding established an important control point in their biosynthesis—the release of substrate. This occurs in response to numerous stimuli that act at the cell surface. Dr. Bengt Samuelsson's extensive study of the metabolism of prostaglandins indicated that they are rapidly inactivated on a single pass through pulmonary circulation. Thus, they cannot act as circulating hormones and appear to be made on demand in or in the vicinity of target tissues leading to the concept that prostaglandins are local hormones or autocoids.

Altered production, qualitative and/or quantitative, of prostaglandins and leukotrienes has been implicated in the development of a number of disease states (e.g., atherosclerosis, inflammatory diseases, asthma). Evidence has been accumulating in the literature suggesting that prostaglandins and leukotrienes may stimulate or inhibit various steps in the complex etiology of cancer, i.e., steps in the progression from a transformed cell to a metastatic tumor. The initial volumes in this series will examine the roles of prostaglandins and leukotrienes in tumor initiation, tumor promotion, tumor cell growth and differentiation, tumor immunity, tumor metastasis and cancer therapy. We hope as this field of cancer research develops that this series, Prostaglandins, Leukotrienes, and Cancer, will provide a forum within the framework of current evidence for the synthesis of new hypotheses and discussion of controversial issues.

Kenneth V. Honn
Lawrence J. Marnett

PREFACE

In reviewing varied topics of scientific investigations in the broad field of carcinogenesis, one cannot help but note the increased interest in tumor promotion and prostaglandins that has developed and the rapid advancements which are taking place in these areas. The importance of both inflammation and cell proliferation in tumor promotion suggested to researchers the critical involvement of prostaglandins (arachidonic acid cascade) in the process of tumor promotion. An apparent marriage of sorts has taken place between these two important areas of investigation with the resulting research providing some notable results that are aiding in the unraveling of the mysteries of cancer.

This volume, "Arachidonic Acid Metabolism and Tumor Promotion," is one of a series covering a wide range of research on the importance or role of arachidonic acid metabolites on the induction and treatment of cancer. Included in this volume are chapters that cover both arachidonic acid metabolism in the skin and the role(s) of arachidonic acid metabolites in skin tumor promotion and promoter-induced inflammation and epidermal hyperplasia. The effects of phorbol ester tumor promoters on phospholipid metabolism *in vitro* are examined and the effects tumor promoting phorbol esters may exert on cell membrane signal transmission by protein kinase C are presented. In another chapter, the relationship between active oxygen or more broadly the prooxygen state and promotion is presented. The possible involvement of arachidonate products in tumor promoter inhibition of cell-cell communication is covered and finally a chapter on proteases and cyclic nucleotides is included because of their effect on and relationship to the arachidonic acid cascade.

It is the general aim of this volume to carefully examine the scientific relationship between arachidonic acid metabolism and tumor promotion. Various inhibitors of arachidonic acid metabolism may very well be useful chemopreventive agents against cancer.

It is indeed hoped that this volume will encourage further scientific investigations and therefore a better understanding of the role of arachidonic acid metabolites in tumor promotion. Of interest to all scientists involved in medical research especially related to cancer, it will also be of special interest to pharmacologists, biochemists, cell biologists, molecular biologists, pathologists, and toxicologists.

INTRODUCTION

S.M. Fischer and T.J. Slaga

The importance of prostaglandins, thromboxanes, prostacyclin, hydroperoxy fatty acids and leukotienes in cancer can no longer be considered simply a possibility; work in recent years has clearly indicated such a reality. These compounds are produced by the oxidation of 20 carbon fatty acids such as arachidonic acid (see Figure 1). The level of involvement of these eicosanoids, the term applied to all arachidonate metabolites, in cancer occurs at all levels from carcinogen metabolism to tumor promotion to the rate of proliferation of tumors to the interaction between the tumor and host to the metastatic potential of tumors. This volume, one of a series intended to cover this spectrum of the disease we call cancer, is devoted to one aspect, tumor promotion.

The transformation of a normal cell to a cancerous one is a complex process which can be induced by any of a variety of reagents, including viruses, radiation and chemicals. Each of these agents can cause an alteration in the genome and/or the rate of expression of the genetic material. This irreversible step is called initiation. Such changes may not be manifested in tumors if the dose or level of initiator is low. Subsequent repeated exposure to noncarcinogenic compounds called promoters, however, results in tumors, the phenotypic expression of the initiation event. One of the best studied models in this regard is the two-stage carcinogenesis model in mouse skin. The promoter most commonly used, and one of the most potent so far identified, is tetradecanoylphorbol acetate (TPA or PMA); its mechanism of action is not completely understood, however. Application of TPA as well as of other promoters to mouse skin triggers numerous morphological and biochemical events including such responses as an increased phospholipid metabolism, an increase in arachidonic acid release and an elevation of prostaglandin levels which are described in detail in this volume. These events may contribute to the intense inflammation produced by TPA. Eicosanoids are further implicated in the promotion process by the findings that inhibitors of various parts of the arachidonic acid cascade can modulate the tumor response. Because several of the same inhibitors effectively inhibit specific TPA-induced events, such as ornithine decarboxylase activity, the role of arachidonate metabolites in promotion is clearly implicated.

The work described in this volume, while primarily using TPA as the prototype promoter and frequently using mouse skin as the model, has far reaching implications. The phenomenon of tumor promotion has been described for other organs (e.g. liver, bladder, esophagus), other types of compounds (e.g. benzoyl peroxide, anthralin, teleocidin, saccharin) and other species, most notably man. The ultimate significance in understanding the mechanism of tumor promotion, as well as any other aspect of the neoplastic process, lies in the possibility that cancer can be prevented by intervening in the complex series of events that occur in carcinogenesis. For this reason the role of archidonate metabolites in physiological and pathological states is under increasingly more intensive investigation.

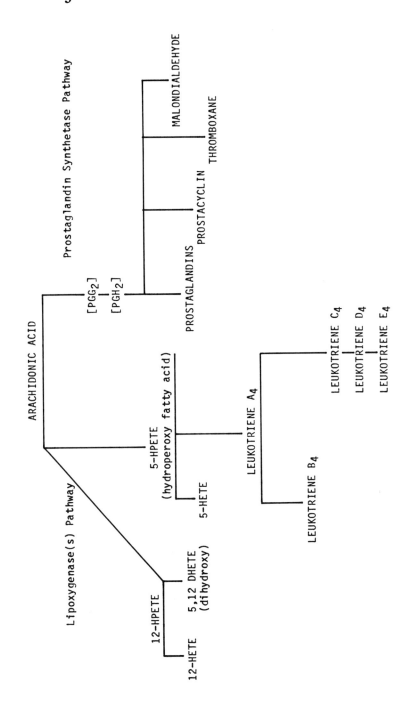

Chapter 1

ARACHIDONIC ACID METABOLISM IN THE SKIN
Vincent A. Ziboh
University of California
Davis, CA 95616

Page

I. Introduction. 6
   A. Essential fatty acids. 6

II. Biosynthesis of eicosanoids from arachidonic acid by skin
    epidermis. 8
   A. Transformation of arachidonic acid via the
      cyclooxygenase pathway. 8
   B. Transformation of arachidonic acid via the lipoxygenase
      pathways. 9

III. Metabolism of prostaglandins by skin epidermis 11

IV. Functions of eicosanoids in the skin 12
   A. Role in normal physiological processes. 12
   B. Altered metabolism of arachidonic acid and cutaneous
      hyperproliferative processes. 12

V. Receptors for arachidonic acid metabolites in the
   epidermis. 14

   References 16

S.M. Fischer and T.J. Slaga (eds.), ARACHIDONIC ACID METABOLISM AND TUMOR
PROMOTION. Copyright © 1985. Martinus Nijhoff Publishing, Boston. All rights reserved.

1. ARACHIDONIC ACID METABOLISM IN THE SKIN
VINCENT A. ZIBOH

I.   INTRODUCTION

The prostaglandins and related arachidonate metabolites are not normally synthesized or stored to any extent in body tissues. Increased levels are brought about by rapid biosynthesis in response to a variety of chemical, physical and hormonal factors. This synthesis or arachidonic acid cascade is complex and includes several enzymes that are affected by endogenous and/or exogenous regulatory factors. Generally, molecular oxygen is inserted into a polyunsaturated fatty acid with acyl chain rearrangement to form the various structures. The principle end products of the various branches of the pathway are the prostaglandins, the thromboxanes, prostacyclin, the hydroperoxy and hydroxyfatty acids, and the leukotrienes. All are derived from the essential fatty acids in the diet.

A.   Essential Fatty Acids:

Although linoleic acid (9,12-octadecadienoic acid) and linolenic acid (9,12,15-octadectrienoic acid) are two major essential fatty acids derived from dietary sources, they are of two metabolically unrelated families in animal metabolism. Linoleic acid (18:2) for instance, undergoes microsomal desaturation and elongation to form arachidonic acid (20:4) which belongs to the family of n6 acids, whereas linolenic acid (18:3) undergoes similar metabolic transformation to form eicosapentaenoic acid (20:3) which belongs to the family of the n3-acids. A schematic diagram of the metabolism of these two dietary fatty acids into their longer chain metabollites is shown in Figure 1.

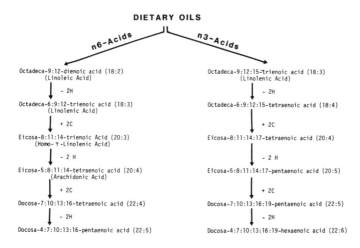

FIGURE 1: Schematic diagram of the in vivo desaturation and elongation of dietary essential fatty acids of the n3 and n6 series.

Enthusiasm for the role of certain polyunsaturated fatty acids (PUFA) in skin biology was heightened after Burr and Burr (1,2) described that rats maintained on a fat-free diet over a long period developed external abnormalities characterized by growth retardation, severe scaly dermatosis of the dorsal and pedal skin, caudal necrosis and extensive water loss through skin. Because the cutaneous symptoms could be reversed by certain dietary polyunsaturated fatty acid (EFA), notably, linoleic acid and arachidonic acid and because these fatty acids are not biosynthesized in the skin in amounts sufficient for regulating physiological processes, considerable interest is now directed towards understanding their mode of biosynthesis, metabolism, and role in the physiology and pathophysiology of the skin.

The occurence of oxidative desaturation and elongation of dietary linoleic acid into arachidonic acid (5,8,11,14 eicosatetraenoic-acid) by skin enzyme preparations remains to be established. Despite this gap in our knowledge of the biosynthesis of these essential fatty acids in the skin, their occurence as esters of phospholipids in this tissue is well described (3,4). It is likely that their accumulation in this class of lipids is to sustain the membrane fluidity of the epidermal cells and to serve as a reservoir for providing arachidonic acid as a substrate for oxidative transformation into potent biological molecules via the cyclooxygenase and lipoxygenase pathways.

II. BIOSYNTHESIS OF EICOSANOIDS FROM ARACHIDONATE BY SKIN EPIDERMIS

  A. Transformations of Arachidonic Acid via the Cyclooxygenase Pathway.

When skin is subjected to a variety of stimuli (mechanical, chemical, or hormonal) precursor arachidonic acid is released predominantly from skin phosphatidylcholine by the activation of a skin phospholipase A_2 (5). Phosphatidylinositol may also contribute arachidonic acid for biosynthesis of the eicosanoids (6). Once released, arachidonic acid undergoes oxidative metabolism into a variety of potent biological products. Figure 2 illustrates the oxidative transformation of arachidonic acid via the cyclooxygenase pathway.

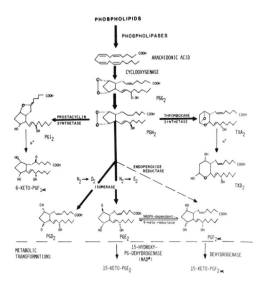

FIGURE 2: Transformations of arachidonic acid released from skin phospholipids into thromboxane, prostacyclin, prostaglandins and their major degradative products.

Biosynthesis of cyclooxygenase products from arachidonic acid has now been well documented in skin preparations from frogs (7), rats (8-10), mice (11), guinea pigs (12) and humans (13-16). The prostaglanding synthetase enzyme system is associated with the epidermal membrane particles and has been localized in the microsomal fraction of human epidermis (14).

Biosynthesis of cyclooxygenase products by cultures of epidermal cells (keratinocytes) have also been reported. For instance, fresh human epidermal cells obtained after trypsinization of human epidermis produce prostaglandins (PGs) E_2 (17). Cultures of neonatal as well as adult murine keratinocytes (16,18) exhibited stimulated synthesis PGE_2 and $PGF_{2\alpha}$ in response to 12-O-tetradecanoyl phorbol-13-acetate (TPA). In these studies, the major arachidonic acid metabolites in epidermis from human as well as from several experimental animals are PGE_2 and $PGF_{2\alpha}$ with minor amounts of PGD_2, TXA_2 and PGI_2. A striking exception is in guinea pig epidermis where PGD_2 was reported to be the major metabolites (12).

B. Transformations of Arachidonic Acid via the Lipoxygenase Pathways

(i) 12-Lipoxygenase Pathyway

Interest in the possible role of lipoxygenase products in the biology of skin was sparked when it was demonstrated that epidermal extracts from skin of psoriatic patients accumulated elevated levels of 12-L hydroxy-5,8,10,14-eiocosatetraenoic acid (12-HETE) when compared to uninvolved psoriatic epidermis or normal epidermis (15). Mouse epidermis, as well as cultured neonatal mouse keratinocytes, have been reported to generate 12-HETE (15) from arachidonic acid.

(ii) 5-Lipoxygenase Pathway

Progress in the elucidation of the chemical structures and biologic properties of novel lipid mediators derived from arachidonic acid which are termed leukotrienes has stimulated great interest and curiosity to determine whether or not the 5-lipoxygenase pathway also operates in cutaneous metabolism of arachidonic acid. Recent reports(19) showed that the challenge of ^{14}C-labeled murine keratinocytes with calcium ionophore (A-23187) resulted in the generation of hydroxy fatty acids which were chromatographically similar to authentic 5-hydroxy-eicosatetraenoic acid (5-HETE) and 5,12-dihydroxy-eicosatetraenoic acid (5,12-DHETE). Interestingly, the synthesis of these lipoxygenase products were enhanced when incubations were carried out in the presence of a cyclooxygenase inhibitor (indomethacin). Furthermore, human cultured

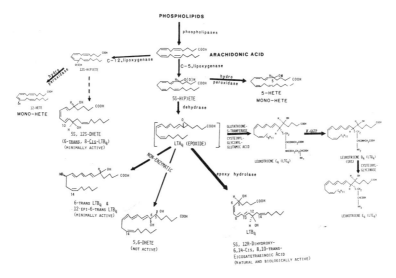

FIGURE 3: Formation of lipoxygenase products from arachidonic acid.

keratinocytes isolated from mastectomy specimens when challenged with calcium ionophore (A-23187) were also reported to generate leukotriene-B-like material from arachidonic acid (20). These observations suggest the existence in the skin of the 5-hydroxy peroxidase and the epoxide hydrolase, two key enzymes which catalyze the formation of 5-HETE and 5,12-DHETE (LTB$_4$) respectively. On the other hand, the possibility that enzymes which catalyze the transformation of LTA$_4$ (an epoxide intermediate in the 5-lipoxygenase pathway) into the peptido-hydroxyeicosatetraenoic acids (LTC$_4$, LTD$_4$ and LTE$_4$) also exist in the epidermis remains to be determined. Figure 3 illustrates the oxidative transformation of arachidonic acid via the 5- and 12-lipoxygenase pathways which have now been confirmed to operate in the epidermis.

More recently increased activity of the 5-lipoxygenase was localized in the high speed supernatant fraction (cytoplasmic) of the lesional epidermis taken from the psoriatic plaque (21). Because of the known infiltration of polymorphonuclear (PMNs) cells into the lesional areas of psoriatic epidermis, it cannot be excluded that some of the 5-lipoxygenase activity observed in these studies may have been contributed by contaminating PMNs. To circumvent this problem, incubations with extracts from the non-lesional epidermis (uninvolved) removed from the psoriatic skin which contains no PMNs demonstrated an increased activity of the 5-lipoxygenase when compared to extracts from non-diseased normal human epidermis. These latter two results indicate first that enzymes which catalyze the transformation of arachidonic acid via the 5-lipoxygenase pathway are present in normal human epidermis. Secondly, this activity is elevated in both the uninvolved as well as the involved psoriatic epidermis. Consistent with these enzymatic studies are reports which showed elevated LTB_4 in fluids derived from abraded psoriatic plaques when compared with abraded uninvolved and normal epidermis (22). More recently, 5-HETE and LTB_4 were detected in psoriatic scales (23).

III. METABOLISM OF PROSTAGLANDINS BY SKIN EPIDERMIS.

A major route for the degradation of PGE_2 and $PGF_{2\alpha}$ in mammalian tissues involves the enzymatic conversion of these primary PGs into 15-keto and 13,14-dihydro-PGs. Activity of the initiating enzyme (15-hydroxyprostaglandin dehydrogenase) has been demonstrated in human skin (13) and in rat skin (24). Furthermore, the conversion of PGE_2 into $PGF_{2\alpha}$ by an NADPH-dependent PGE_2-9-ketoreductase has also been demonstrated in both human and rat skin preparations (25). Interestingly, the activity of this PGE_2-9-ketoreductase was found to be elevated in the lesional epidermis from psoriatic plaque as well as in the hyperproliferative epidermis from essential fatty acid (EFA)-deficient rats. Increase in the activity of the PGE_2-9-ketoreductase is due at least in part to the increased generation of NADPH via the pentose phosphate pathway by the two hyperproliferative epidermal specimens. It is at present not clear whether or not the transformation of PGE_2 into $PGF_{2\alpha}$ is fundamental to cellular proliferation, thus, this important interconversion of PGE_2 and $PGF_{2\alpha}$ remains to be delineated. The existence of this enzyme in the skin nonetheless suggests that pathophysiological and physiological processes in this tissue may be regulated by the relative amounts of the E and F prostaglandins produced in the tissue.

IV. FUNCTIONS OF EICOSANOIDS IN THE SKIN

A. Role in Normal Physiological Processes

The precise roles(s) of the prostaglandins and the recently identified products of the lipoxygenase pathway in normal cutaneous physiology remains unclear. However, PGE_2 has been reported to stimulate the activity of adenyl cyclase, an enzyme which catalyzes the transformation of adenosine triphosphate (ATP) into cyclic adenosine-3',5'-monophosphate (cAMP) in epidermal preparations from humans (26,27), domestic pigs(28), and in guinea pig keratinocytes (29). Since a number of studies suggest that elevation of cAMP levels in mammalian epidermis takes place during maturation (keratinization), it is likely that products of both cyclooxygenase and lipoxygenase pathways are involved, at least in part, in these intracellular processes. However, the precise relationship between prostaglandins, cyclic nucleotides, leukotrienes and epidermal maturation remains to be clarified.

B. Altered Metabolism of Arachidonic Acid and Cutaneous Hyperproliferative Processes

The unique cellular activity of skin is keratinization terminating in the maturation of the epidermal cells. Alteration of this orderly dynamic process results in abnormal epidermal maturation and scaling. Fig 1 demonstrates a typical effect dietary deficiency of essential fatty acids (EFAs) on rats. Thus, the skin of the EFA-deficient rats provides a useful model for studying the interplay between essential fatty acids, prostaglandins, products of cyclooxygenase and lipoxygenase pathways and epidermal proliferation. In earlier studies, prostaglandin derivatives of the E and B series were reported to accelerate maturation of the chick embryo skin, resulting in epidermal thickening, precocious keratinization and failure of downfeather morphogenesis (30,31). Rats fed a diet deficient in essential fatty acids exhibit characteristic scaly lesions as shown in Figure 4. In these animals, epidermal prostaglandins are reduced markedly (9) and homogenates of the skin inhibited the oxidative transformation of arachidonic acid into PGs by sheep vesicular microsomes (32). This inhibitory effect was later found to be due to elevated levels of eicosa-5, 8, 11-trienoic acid (20:3, n9) which accumulated in the epidermis of the EFA-deficient animals. Furthermore, skin from these EFA deficient rats are hyperproliferative and exhibit a germinative hyperplasia leading to an increased transit rate through the skin (33). A similar hyperplasia and elevated DNA synthesis have also been demonstrated in hairless mice fed a diet deficient in EFA (34). These findings suggest

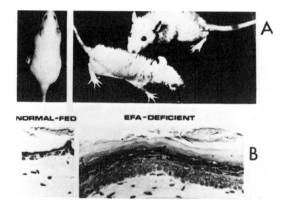

FIGURE 4: Effects of dietary essential fatty acids on rats. (A) represents normal-fed rat (left) and EFA-deficient fed rats (right) (B) represents histological evaluation of normal-fed rat (left) and EFA-deficient fed rat (right).

that altered metabolism of arachidonic acid probably enhanced by the increased biosynthesis of an endogenous inhibitor of cyclooxygenase pathway such as 20:3, n9 fatty acid may be contributory, at least in part, to the development of a scaly dermatosis in the skin of the EFA deficient rats. Indeed, the topical application of the 20:3, n9 fatty acid to skin of hairless mice produced an intense scaly dermatosis which is characterized by hyperplasia and acanthosis of the epidermal layer (35). Also consistent with this concept of the possible existence of an endogenous inhibitor of the cyclooxygenase pathway in the skin, it has been reported that homogenates from psoriatic plaque markedly inhibited the oxygenation of arachidonic acid into PGs by acetone powder of sheep vesicular gland (36). In another report, Hammarstrom et al. (15) found a marked accumulation of precursor arachidonic acid as well as 12-HETE in lesional epidermis from psoriatic plaque, implying a redirection of arachidonic acid metabolism towards the lipoxygenase pathway. Consistent with this view are recent reports of increased activity of the 5-lipoxygenase pathway in extracts from lesional areas of psoriatic epidermis as well as elevated levels of lipoxygenase products such as LTB_4 and 5-HETE which have been detected in psoriatic scales as well as in fluid derived from abraded psoriatic plaques (22,23).

V. RECEPTORS FOR ARACHIDONIC ACID METABOLITES IN THE EPIDERMIS

Evidence for the existence of prostaglandin receptors in a variety of mammalian tissues and the association of this ligand-receptor interaction with specific biological actions of the PGs (37-40) prompted investigations to determine whether or not receptors for PGs exist in the epidermis. Receptors for PGE_2 and $PGF_{2\alpha}$ were first demonstrated in membrane preparations from normal-fed and EFA-deficient rats(41). The concentration of the binding sites of $PGF_{2\alpha}$ increased 5-fold in the membrane preparations from the hyperproliferative skin of the EFA-deficiency rats. To further explore the ligand-receptor interactions of PGE_2 and $PGF_{2\alpha}$ during the EFA-deficiency, specific binding of PGE_2 and $PGF_{2\alpha}$ to skin membrane preparations were carried out at two weekly intervals during the feeding of EFA-deficient diet to rats (42). Results in Figure 5 demonstrated that a relationship exists between increased binding capacity of the membrane preparations for $PGF_{2\alpha}$ and the increased development of epidermal hyperplasia and proliferation of the EFA-deficient rats. These findings indicate that ligand-receptor interactions are altered during the development of hyperproliferation in the epidermis. It remains to be established, however, whether the

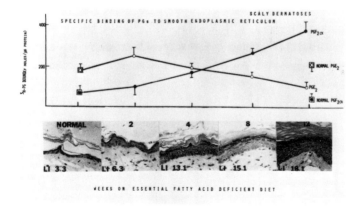

FIGURE 5: Time-course of altered specific binding of ^{3}H-PGE$_2$ and ^{3}H-PGF$_2$ to epidermal membrane preparations from skin of normal-fed and EFA-deficient rats. [Upper Chart] represents specific binding of PGs to membrane preparations [Lower Chart] represents photomicrographs from normal (0) through EFA-deficient (2-12). LI (labeling index) represents the percent of germinative cells in DNA synthesis (S-phase).

increased epidermal binding for PGF$_{2\alpha}$ implicates this prostanoid directly with the hyperproliferative activity in the epidermis of the EFA-deficient rats. Two reported observations have suggested such a direct relationship between PGF$_{2\alpha}$ and hyperproliferation. In one study, ^{3}H-PGF$_{2\alpha}$ was reported to bind strongly to nuclear chromatin in the epidermal neoplastic cells (43) and suggesting a role for PGF$_{2\alpha}$ in DNA synthesis and cutaneous carcinogenesis. In another study, PGF$_{2\alpha}$ at low concentrations was shown to stimulate DNA synthesis in quiescent Swiss mouse 3T3 cells in culture (44) whereas high concentrations inhibited it. These two reported observations suggest a possible role for PGF$_{2\alpha}$ in the regulation of cell proliferation.

Receptor for PGE$_2$ has also been demonstrated in membrane preparations from human skin(45). A Scatchard plot of the equilibrium data from such a study is shown in Figure 6. The insert diagram shows binding site concentration of 210 x 10^{-12}M. Exposure of the human skin epidermal membrane preparations to UVB-irradiation prior to binding experiments resulted in the loss of the membrane binding capacity of PGE$_2$. The UVB-induced inhibitory effect can be prevented by 5,5'-dithiobis-(2-

nitrobenzoic acid), a known protein sulfhydryl-oxidizing agent. The results suggest that UVB-irradiation possibly initiate the reduction of critical protein disulfide groups and the peroxidation of lipids in the membranes which are essential for the receptor-PGE_2 interaction.

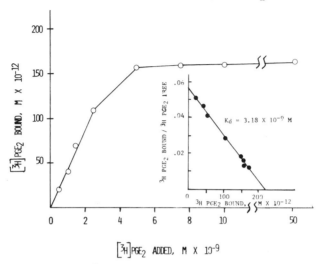

FIGURE 6: Binding of ^3H-PGE_2 to human epidermal membrane preparations and an insert representing the Scatchard plot of the binding data.

REFERENCES

1. Burr GO, Burr MM: A new deficiency disease produced by the rigid exclusion of fat from the diet. J Biol Chem (82) 345-367, 1929.
2. Burr GO, Burr MM: On the nature and role of fatty acids essential in nutrition, J Biol Chem (86) 587-621, 1930.
3. Vroman HE, Nemeek RA and Hsia SL: Synthesis of lipids from acetate by human preputial and abnominal skin in vitro. J Lipid Res (10) 507-514, 1969.
4. Gray GM and Yardley HJ: Lipid composition of cells isolated from pig, human and rat epidermis. J Lipid Res (16) 434-440, 1975.
5. Ziboh VA and Lord JT: Phospholipase A activity in the skin: Modulators of arachidonic acid release from phosphatidylcholine. Biochem J (184) 283-290, 1979.

6. Galey CI, Ziboh VA, Marcelo CL and Voorhees JJ: Activation of phospholipase A and C in murine keratinoctyes by phorbol ester 12-O-tetradecanoylphorbol-13-acetate. J Invest Dermatol (80) 345, 1983.
7. Jessup SJ, McDonald-Gibson WJ, Ramwell PJ and Shaw JE: Biosynthesis and release of prostaglandins on hormonal treatment of frog skin and their effect on ion transport. Fed Proceedings (29) 387, 1970.
8. Ziboh VA and Hsis SL: Prostaglandin E_2: Biosynthesis and effects on glucose and lipid metabolism in rat skin. Arch Biochem Biophys (146) 100-109, 1971.
9. Van Dorp D: Recent developments in the biosynthesis and the analysis of prostaglandins. Annals of the New York Academy of Science (180) 181-199, 1971.
10. Greaves MW and McDonald-Gibson W: Effect of non-steroid anti-inflammatory drugs on prostaglandin biosynthesis by skin. Brit J Dermatol (88) 47-50, 1973.
11. Wilkinson DI and Walsh JT: Prostaglandin biosynthesis in the epidermis and dermis of young mouse skin and the effects on calcium and cyclic nucleotides. J Invest Dermatol (68) 210-214, 1977.
12. Ruzicka T and Printz MP: Arachidonic acid metabolism in guinea pig skin. Biochem Biophys Acta (711) 391-397, 1982.
13. Jonsson DC, Angaard E: Biosynthesis and metabolism of prostaglandin E_2 in human skin. Scan J Clin Lab Invest (29) 289-296, 1972.
14. Ziboh VA: Biosynthesis of prostaglandin E_2 in human skin: Subcellular localization and inhibition of unsaturated fatty acids and anti-inflammatory drugs. J Lipid Res (14) 377-384, 1973.
15. Hammarstrom S, Hamberg M, Samuelsson B, Duell EA, Strawiski M, Voorhees JJ: Increased concentration of non-esterified arachidonic acid. 12L-hydroxy-5,8,10,12-eicosatetraenoic acid, prostaglandin E_2 and prostaglandin $F_{2\alpha}$ in epidermis of psoriasis. Proc Nat Acad Sci (72) 5130-5132, 1975.
16. Hammarstrom S, Lingren JA, Marcelo C, Duell EA, Anderson RF and Voorhees JJ: Arachidonic acid transformations in normal and psoriatic skin. J Invest Dermatol (73) 180-183, 1979.
17. Forstrom L, Goldyne ME and Winklemann RK: Prostaglandin production by human epidermal cells in vitro: A model for studying pharmacolo-

gical inhibition of prostaglandin synthesis. Prostaglandins (8) 107-113, 1974.

18. Fischer SM and Slaga TJ: Modulation of prostaglandin synthesis and tumor promotion. In: Proc. Internatl. Cong. Prostaglandins and cancer. Alan R. Liss, N.Y., 1982, pp 255-264.

19. Ziboh VA, Marcelo C and Voorhees JJ: Induced lipoxygenation of arachidonic acid in mouse epidermal keratinocytes by calcium ionophore, A23187. J Invest Dermatol (76) 307, 1981.

20. Brian SD, Camp RDR, Leigh IM and Fordtlutchinson AW: The synthesis of leukotriene B_4-like material by cultured human keratinocytes. J Invest Dermatol (78) 328, 1982.

21. Ziboh VA, Casebolt T, Marcelo CL and Voorhees JJ: Enhancement of 5-lipoxygenase activity in soluble preparations of human psoriatic plaque preparation. J Invest Dermatol (80) 359, 1983.

22. Brain S, Camp R, Dowd P, Black AK, Woolard PM, Mallet AI and Greaves MW: Psoriasis and Leukotriene B_4. Lancet (2) 762-763, 1982.

23. Camp R, Mallet AI, Woolard PM, Brain S, Black AK and Greaves MW: Monohydroxy metabolites of arachidonic acid and linoleic acids in psoriatic skin. J Invest Dermatol (80) 359, 1983.

24. Camp RD and Greaves MW: The catabolism of prostaglandins by rat skin. Biochem J (186) 153-160, 1980.

25. Ziboh VA, Lord JT and Penneys NS: Alterations of prostaglandin E_2-9-ketoreductase activity in proliferating skin. J Lipid Res (18) 37-43, 1977.

26. Mui MM, Hsia SL and Halprin KM: Further studies on adenyl cyclase in psoriasis. Brit J Dermatol (92) 255-262, 1975.

27. Adachi K, Yoshikawa K, Halprin KM and Levine U: Prostaglandin and cAMP in epidermis. Evidence for the independent action of prostaglandins and adrenaline on the adenyl cyclase system of pig and human epidermis, normal and psoriatic. Brit J Dermatol (92) 381-388, 1975.

28. Ilzuka H, Adachi K, Halprin KM and Levine V: Adenosine and adenine nucleotides stimulation of skin (epidermal) adenyl cyclase. Biochem Biophys Acta (444) 685-693, 1976.

29. Wilkinson DI and Orenberg EK: Effect of prostaglandins on cyclic nucleotide levels in cultured keratinocytes. Prostaglandins (17) 419-429, 1979.

30. Kirscher CW: Effects of specific prostaglandins on development of chick embryo skin and down feather organ in vitro. Develop Biol (16) 203-215, 1967.
31. Kirscher CW: Accelerated maturation of chick embryo skin treated with a prostaglandin - 278 (PGB): An electron microscopy study. Am J Anat (124) 491-512, 1969.
32. Ziboh VA, Vanderhoek JY, Lands EM: Inhibition of sheep vesicular gland oxygenase by unsaturated fatty acids from skin of essential fatty acid deficient rats. Prostaglandins (5) 233-239, 1974.
33. McCullough JL, Schreiber SH and Ziboh VA: Cell proliferation kinetics of epidermis in the essential fatty acid deficient rat. J Invest Dermatol (70) 318-320, 1978.
34. Lowe NJ and Stoughton RB: Essential fatty acid deficient hairless mouse: A model of chronic epidermal hyperproliferation. Brit J Dermatol (96) 155-162, 1977.
35. Nguyen TT, Ziboh VA, Uematsu S, McCullough JL and Weinstein G: New Model of Scaling Dermatosis: Induction of hyperproliferation in hairless mice with eicosa-5,8,11-trienoic acid. J Invest Dermatol (76) 384-387, 1981.
36. Pennys NS, Ziboh VA, Lord JT and Simon P: An inhibitor of prostaglandin synthesis in psoriatic plaque. Nature (254) 351-352, 1975.
37. Kuehl FA, Humes JL, Tarnoff J, Cirillo VJ and Ham EA: Prostaglandin receptor site: Evidence for an essential role in the action of luteinizing hormone. Science (169) 883-886, 1970.
38. Saez JM, Dazord A and Gallet D: ACTH and prostaglandin in human adrenocortical tumors. Apparent modification of a specific component of the ACTH-binding site. J Clin Invest (56) 536-547, 1975.
39. Rao CV: Receptors of prostaglandins and gonadotropins in cell membranes of bovine corpus luteum. Prostaglandins (4) 567-576, 1973.
40. Lefkowitz RJ, Mullikin D, Wood CL, Gore TB, Mukherjee C: Regulation of prostaglandin receptors by prostaglandins and guanine nucleotides in frog erythrocytes. J Biol Chem (252) 5295-5303, 1977.
41. Lord JT, Ziboh VA and Warren SK: Specific binding of prostaglandins E_2 and F_α by membrane preparations from rat skin. Endocrinology (102) 1300-1309, 1978.
42. Lord JT, Ziboh VA, McCullough JL: Alteration of specific binding of prostaglandin F_2 by hyperproliferative skin membrane fractions

from essential fatty acid-deficient rats. Prostaglandins and Medicine (2) 21-31, 1979.

43. Lupulescu AP: Autoradiographic studies of ^{3}H-Prostaglandin $F_{2\alpha}$ into nuclei of epidermal neoplastic cells. J Invest Dermatol (72) 279, 1979.

44. DeAsua, Clingan L and Rudland PS: Initiation of cell proliferation in cultured mouse fibroblasts by prostaglandin $F_{2\alpha}$ Proc Nat Acad Sci (72) 2724-2728, 1975.

45. Lord JT and Ziboh VA: Specific binding of prostaglandin E_2 to membrane preparations from human skin: receptor modulation by UVB-irradiation and chemical agents. J Invest Dermatol (73) 373-377, 1979.

Chapter 2

ARACHIDONIC ACID METABOLISM AND TUMOR PROMOTION
Susan M. Fischer
University of Texas System Cancer Center
Smithville, TX 78957

		Page
I.	Introduction.	22
II.	Tumor promotion studies using prostaglandins or inhibitors	25
III.	Related in vitro studies	
	A. Indomethacin enhancement of HPETE production	28
	B. An in vitro model.	29
	C. Metabolic cooperation.	31
IV.	Possible involvement of free radicals	33
V.	Conclusion.	42
References		42

S.M. Fischer and T.J. Slaga (eds.), ARACHIDONIC ACID METABOLISM AND TUMOR PROMOTION. Copyright © 1985. Martinus Nijhoff Publishing, Boston. All rights reserved.

2. ARACHIDONIC ACID METABOLISM AND TUMOR PROMOTION
S.M. FISCHER

I. INTRODUCTION

Experimental chemical carcinogenesis studies in animals are valuable in identifying those biological events or agents that play either an essential or modulatory role in the development of neoplasias. Mouse skin has proved to be one of the best model systems for studying the multistage nature of carcinogenesis (1-3). Skin tumors in mice can be readily induced by the sequential application of a subthreshold dose of a carcinogen (initiation stage) followed by repetititive treatment with a noncarcinogenic tumor promoter (promotion stage). The initiation stage requires only a single application of either a direct-acting carcinogen or a procarcinogen which requires metabolism before becoming active; it is of an essentially irreversible nature, while the promotion stage is thought to be initially reversible but later becoming irreversible. This second stage, promotion, is most often accomplished by using the phorbol diester 12-0-tetradecanoyl phorbol 13-acetate (TPA), although a variety of agents have been identified as skin tumor promoters, including benzoyl peroxide, anthralin and dihydroteleocidin B (4). In the SENCAR mouse, promotion has been further subdivided into additional stages in which the sequential use of incomplete or partial promoters such as the ionophore A-23187 and mezerein can replace the use of a complete promoter such as TPA (3,5). These stages of promotion have been identified primarily through the use of drugs that were found to inhibit specific biochemical events as well as inhibiting tumor development (5,6).

The mechanism of action of TPA has been under investigation for numerous years and has been hampered in part by the fact that TPA causes a plethora of morphological and biochemical responses in the skin. Of all the observed effects, the induction of three events, epidermal cell proliferation and inflammation, ornithine decarboxylase (ODC) activity, and dark cell appearance, appear to correlate the best with promoting activity (3,7). These correlations do not necessarily hold up for promoters not of the phorbol diester series or with irritating but nonpromoting agents such as ethylphenylpropriolate (4).

Investigations in this laboratory as well as others into the role of prostaglandins and the other arachidonic acid metabolites were prompted in part by the observation that TPA induces cytotoxicity, inflammation, and vascular permeability changes (8,9). The essential nature of inflammation to phorbol diester tumor promotion was suggested by several early studies using the antiinflammatory steroids dexamethasone and fluocinolone acetonide. Fluocinolone acetonide in particular is extrememly potent: Repeated application of as little as 0.01µg almost completely counteracts TPA tumor promotion (6,10). This drug also effectively counteracts TPA induced cell proliferation and inflammation (10). In the time since these studies were done anti-inflammatory steroids have been shown to inhibit the production of arachidonic acid metabolites by inhibiting phospholipase A_2 (11,12). This may be the underlying mechanism by which these agents act as strong inhibitors of TPA-induced inflammation and epidermal DNA synthesis (13,14).

TPA mediated changes in phospholipid metabolism have been noted in a variety of systems and are described in Chapter 4 of this volume. Briefly, Rohrschneider et al (15,16) observed that both skin treated once with TPA and papillomas resulting from repeated treatment of initiated skin have high levels of the phospholipids phosphatidylcholine and phosphatidylethanolamine. Using the lipoxygenase and cyclooxygenase inhibitor 5,8,11,14 eicosatetraynoic acid (ETYA) as well as the cyclooxygenase inhibitor indomethacin, Wertz and Mueller (17) showed that it is the activation of the lipoxygenase pathway that probably plays a key role in TPA induced phospholipid synthesis. Studies by Arffmann and Glavind (18,19) on methyl esters of fatty acids showed that in mice methyl 12-oxo-trans-10-octadecanoate is a weak tumor promoter. This was particularly interesting in light of the hypothesis by Rohrschneider and Boutwell (20) that TPA possibly mimics a natural endogenous product, a likely candidate for which is the family of polyunsaturated fatty acid esters. To support this it was noted that they have striking structural similarities to TPA and also that there is a relationship of the fatty acids of phospholipids and their conversion to the prostaglandins (20). To test this hypothesis, Furstenberger and Marks (21) performed a binding experiment looking for receptor competition. TPA, however, did not compete for the established PGE_1 receptors of the adipose cells.

There is now substantial evidence to demonstrate that TPA provokes prostaglandin (and other arachidonic acid metabolites) synthesis in mouse epidermal cells. In a study by Brune et al (22) comparing a series of tigliane-, ingenane- and daphnane-type diterpene derivatives, it was found that a good correlation existed between irritant and promoting acitivity in mouse skin and PGE_2 release in macrophages. Bresnick et al (23) reported that TPA induced a striking increase in PGE levels as early as 1 hr after treatment of CD-1 mice. This increase reached a peak at 24 hr and remained elevated for at least an additional 24 hr. PGF was also measured but showed only a slight increase at 12 hr. The ratio of PGE/PGF may be more important than the actual levels; under basal conditions the ratio is about 4 but rises to 21 at 48 hr (23). Verma et al (24), however, reported a five-fold increase in $PGF_{2\alpha}$ (as compared to 3.5 for PGE_2) 14 hr after TPA treatment of CD-1 mice. These studies suggested that PGE_2 is the key prostaglandin affected by TPA, although other prostaglandins or arachidonic acid metabolites were not measured.

The induction of ornithine decarboxylase has been suggested as an essential but not sufficient factor for TPA tumor promotion (25). Considerable evidence indicates that prostaglandins are involved in this response. Verma et al (26) demonstrated that pretreatment of CD-1 mice with the cyclooxygenase inhibitor indomethacin markedly inhibited TPA induction of ODC. This inhibitory effect could be reversed by the application of PGE_2 but not $PGF_{2\alpha}$, suggesting that PGE_2 may play a crucial role in TPA-induced elevation of ODC activity. Nakadate et al (27,28) also using CD-1 mice more recently reported that TPA-induced ODC activity was also inhibited by the phospholipase A_2 inhibitors mepacrine or dibromoacetophenone, by the lipoxygenase inhibitor nordihydroguaiaretic acid (NDGA), and by the cyclooxygenase-lipoxygenase inhibitors phenidone and BW755C. Application of PGE_2 to either phenidone or NDGA inhibited ODC did not result in restoration of activity nor did indomethacin further suppress NDGA inhibition of ODC induced by TPA (27). These findings were interpreted as suggesting that not only cyclooxygenase products (i.e., PGE_2) but also lipoxygenase product(s) are involved in the mechanism of ODC induction in mouse epidermis, and a lack of either product results in a failure of ODC induction by TPA (27).

II. TUMOR PROMOTION STUDIES USING PROSTAGLANDINS OR INHIBITORS

Various modifiers of the tumor promotion process have been very useful in our understanding of the mechanism(s) of tumor promotion; drugs related to arachidonic acid metabolism are no exception. The above mentioned studies in particular suggested that tumor promotion might be modified through the application of either exogenous PGs or inhibitors of various parts of the arachidonate cascade. Our first approach to this question, made several years ago, was to determine the effect of individual PGs when applied topically alone or with TPA on initiated mouse skin. This series of experiments, summarized in Table 1, demonstrated that the effect of the various PGs on tumor production depends on both the particular agent used and the time of application with respect to TPA. Specifically, $PGF_{2\alpha}$ enhances TPA tumor promotion in the SENCAR mouse by up to 60% with 10 µg applications while PGE_1 reproducibly inhibits promotion at doses as low as 1 µg (29). None of the PGs tested had promoting activity when applied repetitively to initiated mice (29). Verma et al (24) confirmed this with PGE_2 in CD-1 mice as did Furstenberger and Marks (30) with PGE_2 and $PGF_{2\alpha}$ in NMRI mice.

Somewhat perplexing results come from the studies using arachidonic acid cotreatment with TPA. Since TPA application alone should result in the liberation of free arachidonic acid, it was anticipated that either arachidonic acid alone might cause promotion and/or that enhanced TPA promotion would occur. The results indicated, however, that arachidonic acid had little effect at lower doses but showed dramatic inhibition at doses over 100 µg. On the other hand linoleic acid at the same doses had no effect on TPA-induced tumors (see Table below), suggesting that arachidonic acid is active not because of its fatty acid nature but perhaps because it is a direct precursor to the prostaglandins and lipoxygenase products.

Table 1. Effects of Topical Application of Several Prostaglandins and Precursors on TPA Tumor Promotion in SENCAR Mice.

Prostaglandin	dose (µg)	Effect[a]
PGE_1	>1	▼
PGE_2	10	▲▼
$PGF_{2\alpha}$	1-10	▲
arachidonic acid	>10	▼
linoleic acid	10-500	—

[a]Effect is evaluated as enhancement (▲), inhibition (▼), or no effect (—) on the mean number of papillomas per mouse. Data from ref. 29.
[b]Inhibition occurs when PGE_2 applied within 45 min of TPA application; application 90 min before TPA results in a slight enhancement.

While these studies indicated that PGs could be used to modify tumor yield, they were complicated by the fact that TPA itself induces PG synthesis. The effect of exogenously added PGs on arachidonic acid metabolism in this system is not known. Therefore, the next series of studies involved the use of inhibitors of various parts of the arachidonate cascade. Based on the role of PGs in inflammation, it was originally thought that the nonsteroidal anti-inflammatory agents, ie., the cyclooxygenase inhibitors indomethacin or flurbiprofen, should inhibit promotion. In the SENCAR mouse, however, it was found that topical application of indomethacin can enhance TPA tumor promotion at doses of 25 to 100µg while inhibiting at higher, probably toxic doses (31). This enhancing effect is even more pronounced using the multistage promotion protocol established by Slaga et al (2,5) in which four applications of TPA (first stage) are followed by repetitive applications of the weak promoter mezerein (second stage). When indomethacin is applied 2 hr before TPA in the first stage, the tumor number is double that of TPA alone (32). This enhancement occurred in spite of the fact that indomethacin inhibited ODC induction by TPA (data not shown). Indomethacin has only a mildly enhancing effect on the second stage of promotion. Histological studies have shown that in SENCAR mice indomethacin does not inhibit TPA induced hyperplasia, inflammation or dark cell number (31).

Work by other investigators using other strains of mice have shown a different response to indomethacin. Verma et al (26), using CD-1 mice demonstrated that indomethacin could inhibit the TPA induction of ODC activity and further that PGE_1 and PGE_2 but not $PGF_{2\alpha}$ could overcome this inhibition. Later he showed in addition that in the CD-1 mouse indomethacin inhibited rather than enhanced TPA promotion (24). Furstenberger et al (30) also found inhibition of promotion with indomethacin in the NMRI mouse; the involvement of the PGs in TPA induced cell proliferation and inflammation in this mouse will be discussed in detail in the following chapter. The reasons for the differences between strains or stocks of mice with regard to response to indomethacin is not known; it may be related to the lipoxygenase part of the arachidonate pathway, some components of which are much more potent than the PGs with respect to inflammation.

Another series of tumor experiments were done in our laboratory using inhibitors that are effective in blocking both the cyclooxygenase and lipoxygenase pathways. These studies, summarized in Table 2, suggested that while cyclooxygenase inhibitors enhance promotion, inhibitors of both cyclooxygenase and lipoxygenase inhibit promotion (33). Specifically, both phenidone (1-phenyl-3-pyrazolidinone) and ETYA inhibited promotion up to 45%. Additionally, inhibitors of phospholipase A_2, ie, dibromoacetophenone and the anti-inflammatory steroids, are very potent inhibitors of tumor promotion (33). The work of Nakadate et al (34) and Kato et al (35) have also demonstrated the inhibition of TPA induced tumor promotion in CD-1 mice with the lipoxygenase inhibitors nordihydroguaiaretic acid and quercetin as well as with the phospholipase A_2 inhibitor dibromoacetophenone. In these studies inhibition was not seen with phenidone, although this drug was shown to inhibit TPA-induced ODC activity (34).

Table 2. Effects of Inhibitors of Arachidonate Metabolism on TPA Tumor Promotion

Inhibitor	Dose µg	% TPA Group
Indomethacin	10	138
	25	210
	50	167
	100	108
	200	90
Flurbiprofen	10	130
ETYA	100	45
Phenidone	100	45
Dibromoacetophenone	100	33
	250	25
	500	15

Each group consisted of 30 mice initiated with either 10 or 100 nmol (2.56 µg) dimethylbenz[a]anthracene (DMBA) and promoted with TPA 2µg twice weekly starting one week after initiation. Unless otherwise indicated, the inhibitors were applied concomitantly with TPA. The data is expressed as a percentage of the number of papillomas in the experimental group as compared to the TPA alone control group. Data from ref. 33.

III. RELATED IN VITRO STUDIES

A. INDOMETHACIN ENHANCEMENT OF HPETE PRODUCTION

The results of the inhibitor studies in the SENCAR mouse suggested not only that arachidonic acid metabolites are important and perhaps even essential to the tumor promotion process, but also that the most important metabolites may not be the prostaglandins themselves. It has been shown by Hamberg (36) that inhibition of prostaglandin synthesis results in elevated levels of the hydroperoxy- and hydroxy-eicosatetraenoic acid (HPETE and HETE) in human platelets. Since these products may be important in cell proliferation, as indicated by the high levels found in the hyper-proliferative disease psoriasis (37), they may be responsible, at least in part, for the hyperplasia and inflammation seen with TPA treatments. It was of interest, therefore, to determine whether indomethacin could cause an elevation of HPETEs in TPA treated mouse skin. This was accomplished by using primary cultures of adult mouse epidermal cells prelabeled with ^{14}C-arachidonic acid, for a time shown to give maximal incorporation into the phospholipids, and then exposing the cultures to fresh media containing 1 μg/ml TPA with and without 35.8 μg/ml (10^{-4} M) indomethacin for appropriate time periods. The acidified media was extracted twice with ethyl acetate, and small volumes of the dried extract applied to thin layer chromatography plates which were then developed in the "A-9" solvent system (38). Arachidonic acid, various prostaglandins and a mixture of the HPETEs (kindly provided by L. J. Marnett) were used as reference compounds. After the standards were visualized by iodine vapors, the appropriate corresponding sample zones were scraped for scintillation counting. The results (39) showed that the HPETEs are elevated about 45% over control with TPA alone while TPA + indomethacin give an increase over control of 88%. These results indicated that indomethacin may shunt arachidonic acid into the lipoxygenase pathway, further suggesting that the HPETEs are intimately involved in tumor promotion.

B. AN IN VITRO MODEL

Since cell culture systems have the advantage over many in vivo systems of being easier to manipulate and control, it was felt that an in vitro model for particular events associated with tumor promotion would be of value in determining at least some of the mechanisms involved in promotion (40,41). In this study, primary mouse epidermal cell cultures were used to determine (a) the effect of inhibitors or arachidonate metabolism on TPA-induced DNA synthesis and (b) the effect of these inhibitors on PGE_2 and HPETE synthesis. These results were then correlated with in vivo DNA synthesis studies and tumor promotion experiments employing the same inhibitors.

Primary cultures of SENCAR newborn mouse epidermis were used for these experiments; the culture methods have been previously described (42). For the in vitro DNA synthesis studies, the cultures were treated with the appropriate drugs 24 hrs after plating and then pulse-labelled for 1 hr with [^{3}H] thymidine 5 days later during the previously established optimum response time to TPA. The data for the *in vitro* DNA synthesis studies are shown in Table 3. Both the cyclooxygenase inhibitors indomethacin and flurbiprofen enhanced the rate of TPA-induced DNA synthesis; this enhancement correlates well with the increased number of tumors seen in corresponding tumor experiments. Furthermore, the use of the inhibitors of both cyclooxygenase and lipoxygenase, ETYA and phenidone, resulted in a dose-response inhibition of TPA-induced DNA synthesis, as did the phospholipase A_2 inhibitor dibromoacetophenone. This also correlates well with the inhibition of tumor promotion by these same inhibitors.

In vivo DNA synthesis experiments were carried out on adult mice treated in essentially the same manner as was done for the tumor experiments. TPA induced DNA synthesis maximally at 24 hrs; the effects of the inhibitors were therefore measured at this time by 30 min labelling with [^{3}H] thymidine injected i.p. The results of this study (Table 3) were essentially the same as for the *in vitro* DNA synthesis studies.

It was desirable to determine whether these inhibitors were in fact inhibiting or enhancing either the cyclooxygenase or lipoxygenase pathways. As previously described, this was done by thin layer chromatography on [^{14}-C] arachidonate prelabelled cells. The results of this study (data not shown) indicate that in this culture system indomethacin is a cyclooxygenase but not lipoxygenase inhibitor, while ETYA and phenidone appear to inhibit both pathways. Table 4 provides a summary of the correlation between the *in vitro* and *in vivo* systems.

This *in vitro* system appears, therefore, to be a good model for particular events associated with *in vivo* tumor promoter treatment, which is advantageous since cultures are easier to manipulate and provide a 'sink' for the excretion of arachidonate metabolites. Additionally, this information strengthens the hypothesis that the lipoxygenase pathway is important if not essential in the tumor promotion process.

The question of which lipoxygenase products are synthesized and responsible for the observed phenomena is currently under study. Additionally, preliminary radioimmunoassay experiments done in collaboration with A. Welton of Hoffmann-LaRoche utilizing these epidermal

Table 3. The Effects of Arachidonate Inhibitors on TPA-induced DNA Synthesis in SENCAR Mouse Epidermal Cells In Vitro and In Vivo.

drugs	In Vitro dose µg/ml	% TPA	In Vivo dose µg	% TPA
A. Cyclooxygenase Inhibitors				
Flurbiprofen	1	110	N.D.[a]	N.D.
	10	128[b]		
	25	155[b]		
Indomethacin	3.6	116	25	153
	36.0	130[b]		
B. Cyclo- and Lipoxygenase Inhibitors				
ETYA	1	36[b]	100	81
	10	47[b]	250	68
	25	19[b]		
Phenidone	1	83	100	83
	10	57[b]		
	25	30[b]		
C. Phospholipase A_2 Inhibitor				
Dibromoacetophenone	1	55[b]	500	46
	10	23[b]		

[a]N.D.; not determined
[b]Statistically significant p<.05, students t test

cultures have indicated that TPA induces the production of the leukotrienes.

C. METABOLIC COOPERATION

Of the many biochemical effects elicited by TPA, one of the more interesting changes is that of inhibition of metabolic cooperation as seen in the cell-cell communication assay described by Yotti et al

Table 4 Summary of Effects of Arachidonate Inhibitors on Mouse Epidermis *In Vitro* and *In Vivo*

Inhibitors	In Vitro			In Vivo	
	DNA Synthesis	PGE$_2$	HPETE	DNA Synthesis	Tumor Promotion
A. Cyclooxygenase					
1. Flurbiprofen	↓↓	N.D.	N.D.	N.D.	↓↓
2. Indomethacin	↓↓	↓	↓	↓	↓↓
B. Cyclo- and Lipoxygenase					
1. ETYA	↓↓	↓↓	N.D.	↓↓	↓↓
2. Phenidone	↓↓	↓↓	↓	↓↓	↓↓
C. Phospholipase A$_2$					
1. Steroids	↓	N.D.	N.D.	↓↓	↓↓
2. Dibromoacetophenone	↓	N.D.	N.D.		

N.D. = not determined

These experiments are described in the text.

(43) and Murray and Fitzgerald (44). In this assay, wild-type Chinese hamster V-79 cells (6-thioguanine sensitive) reduce the number of colonies formed by 6-thioguanine resistant cells through a form of intracellular communication (43). TPA effectively inhibits this communication as do other promoters including benzoyl peroxide (45). Since it was suggested (45) that benzoyl peroxide promotion may result from membrane changes caused by free radicals, the possibility that the HPETEs might also inhibit metabolic cooperation was tested. A mixture of HPETEs (supplied by L.J. Marnett) was found to be positive (46) in inhibiting cell communication which indicates that they may have weak tumor promoting activity. However, the slightly positive results (46) given by the same prostaglandins previously demonstrated to lack promoting activity by themselves makes the situation less than clear. It remains to be demonstrated whether the HPETEs have tumor promoting activity in whole animal experiments. The role of metabolic communication in tumor promotion is discussed in detail in Chapter 7 of this volume.

IV. POSSIBLE INVOLVEMENT OF FREE RADICALS

A role for activated oxygen species in tumor promotion has been investigated recently through a variety of different approaches. Work by Slaga et al (45) showed that such free radical generating compounds as benzoyl peroxide, lauroyl peroxide and chloroperbenzoic acid have complete tumor promoting ability. Recently Slaga and coworkers (unpublished data) found that butylated hydroxyanisole and butylated hydroxytoluene are effective inhibitors of TPA and benzoyl peroxide promotion. The superoxide dismutase (SOD) mimetic Cu(II)(3,4-diisopropylsalicylic acid)2 (CuDIPS) has also been shown to inhibit both TPA promotion (47) and TPA-induced ODC (48). The involvement of free radicals in promotion has been suggested by Goldstein et al (49,50) who showed that phorbol ester tumor promoters stimulate superoxide anion radical production in human polymorphonuclear leukocytes as measured by cytochrome c reduction. In this system the tumor promoter induced-free radicals can be inhibited by such inhibitors of promotion as protease inhibitors, retinoids and anti-inflammatory steroids (49). Another means of measuring the effect of TPA on human PMN utilization of oxygen is by measuring the photoemission or chemiluminescence that is accompanied by the generation and energy dissipation of superoxide anion and singlet oxygen (51). Several mechanisms by which TPA treatment of PMNs can result in the generation of excited oxygen species have been suggested and are shown in Figure 1. The degranulation reaction of PMNs is part of their normal response to a variety of foreign materials (50) and may result in extracellular singlet oxygen production (51). The normal oxidative metabolism and electron transport machinery also provide sources of active oxygen; TPA has been shown to enhance such oxidation metabolism (50). Thirdly, peroxidation reactions of lipids, including the intermediates in PG synthesis and the HPETEs are also contributors (51). It has been shown by Kensler and Trush (51) in PMNs that TPA-induced chemiluminescence is dose-responsive and that the activities of a series of phorbol esters correspond with their relative activities as tumor promoters.

Figure 1. Possible Mechanisms for the Generation of Chemiluminescence by TPA in Human PMNs. Chemiluminescence occurs as a result of the emission of light when electronically excited states return to ground state. Scheme derived from ref. 51.

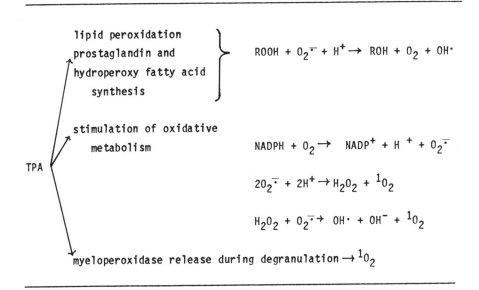

We have recently been using human PMNs to determine whether some of the inhibitors of arachidonic acid metabolism would inhibit TPA-induced chemiluminescence. Measurements were made on samples of 12

$\times 10^6$ freshly isolated PMNs in 3 mls of 0.1% glucose in Dulbecco's phosphate buffered saline using an ambient temperature scintillation counter in the out-of-coincidence mode (50). Figure 2 shows a typical chemiluminescence response to TPA treatment; there is an immediate peak (at about 5 min) followed by a more lengthy return to control levels (more than 30 min). The cells are totally refractory to subsequent TPA treatments. Since the arachidonate cascade can theoretically contribute to the pool of reactive oxygen species it was of interest to determine whether arachidonate inhibitors would have significant inhibitory effect. As indicated in Table 5, all the inhibitors studied to date have considerably reduced the chemiluminescence response of TPA, which suggests that at least part of the reactive oxygen generated is from the arachidonate cascade. Additional work is needed to verify this hypothesis, although it is interesting to note that the addition of arachidonic acid resulted in a chemiluminescence response kinetically similar to TPA (data not shown).

Figure 2. Effects of Indomethacin on TPA-induced Chemiluminescence in Human PMNs.

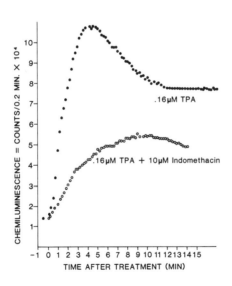

Chemiluminescence was measured after the addition of either TPA (●) or TPA plus indomethacin (o) to 12×10^6 PMNs.

Table 5. Inhibitory Effect on TPA-induced Chemiluminescence in Human PMNs by Arachidonate Inhibitors

Agent	Dose (μM)	% Inhibition of TPA Response
acetone	-	0
indomethacin	100	68
	10	52
phenidone	100	84
	10	66
	1	47
flurbiprofen	100	64
	10	66
dexamethasone	100	42
	10	35

Chemiluminescence was measured after the addition of 160 nM TPA and appropriate dose of drug to vials of 12×10^6 human PMNs. Inhibition was determined as a percent of the TPA alone response.

These studies raised the question of whether TPA could generate active oxygen species in mouse epidermal cells. We recently have modified the chemiluminescence assay for making such determinations (52). The chemiluminescence (CL) enhancer luminol is required in the assay for epidermal cells but not for PMNs, suggesting that the level of oxygen radicals produced is considerably less than in PMNs. This difference may be due to the degranulation and release of myeloperoxidase that occurs in PMNs but not epidermal cells. Using newborn epidermal cells isolated from SENCAR mice, the CL response was shown to be TPA dose dependent and cell number dependent, as indicated in Figure 3. The kinetics of the response (Figure 4) of epidermal cells is such that a distinct rise is seen within 5 min and a peak is reached by 15 min (as compared to a peak response time of 5 min in PMNs). A comparison was made of the abilities of a series of phorbol esters of varying tumor promoting activities to generate CL responses at doses equimolar to TPA. As shown in Table 6, TPA is the most active of the phorbol esters used. Mezerein, a related diterpene that is a weak complete promoter but strong second stage promoter in SENCAR mice (2,5), has at least the same capacity as TPA to stimulate CL.

Figure 3. Effect of TPA Dose and of Cell Number on Epidermal Cell Chemiluminescence.

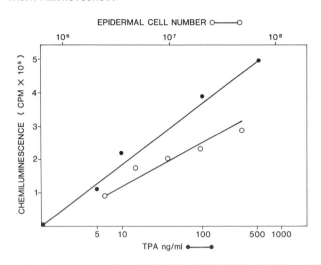

The values represent those at the peak of the response. (o-o), 10^7 cells were used per assay with TPA added at the doses shown. (●-●); the TPA was held constant and the cell number varied.

One of the major oxygen radical species thought to be produced upon TPA activation is the superoxide anion (53). Its participation is indicated through the inhibition seen with the use of SOD, which is specific for the dismutation of this radical. As indicated in Table 7, inclusion of either SOD or the SOD mimetic CuDIPS strongly inhibited the TPA-induced CL response, while catalase, which breaks down H_2O_2, and mannitol, a scavenger for hydroxy radicals, had little effect.

The insertion of molecular oxygen into arachidonic acid and subsequent rearrangements and metabolism produces reactive oxygen species as well as the hydroperoxides of the fatty acid. Because epidermal cells have been shown to produce prostaglandins and other arachidonic acid metabolites in response to TPA, inhibitors of various parts of this metabolic cascade might be expected to affect the levels of radicals

Figure 4. Kinetics of TPA Induced Chemiluminescence in Mouse Epidermal Cells

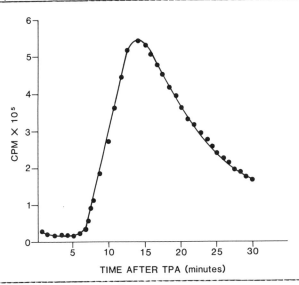

TPA (100 ng/ml) was added to 10^7 mouse epidermal cells and the response monitored continuously.

produced by TPA stimulation. As indicated in Table 8, the effect of the various inhibitors on TPA-stimulated chemiluminescence depends on the particular pathway of arachidonic acid metabolism against which they are effective. Those inhibitors which are predominantly lipoxygenese inhibitors (NDGA and benoxaprofen) or inhibit both cyclooxygenase and lipoxygenese (phenidone and ETYA) are effective in diminishing the response. Cyclooxygenase inhibitors (indomethacin and flurbiprofen), however, do not inhibit at lower doses (10μM or less); at 100 μM some inhibition does occur. This data correlates well with both the effect of these inhibitors on TPA-stimulated DNA synthesis in vitro and with their effects on TPA tumor promotion. The demonstration that inhibitors of arachidonic acid, particularly by lipoxygenase(s), diminish the TPA induced CL response suggests that a major source of the oxygen radicals produced in epidermal cells is through the metabolism of arachidonic acid.

Table 6. Structure-function Studies on the Ability of Tumor Promoters and Related Compounds to Induce Chemiluminescence in Mouse Epidermal Cells[a]

Compound	Dose (nM)	% TPA (162 nM) (peak CL)	Relative Promoting Ability
TPA	162	100	++++
phorbol 12,13 dibenzoate	16.2	0	
	162	13	++
	1620	69	
phorbol 13,20 diacetate	162	0	±
phorbol	162	0	-
mezerein	162	100	±
	1620	303	

[a]10^7 murine epidermal cells were used per 3 ml assay; the cpm at the peak of the response were used to determine % of the TPA peak.

These findings, that TPA can generate oxygen free radicals in mouse epidermal cells, raise several very interesting questions: (I) It may not be necessary to include leukocytes in a theory on the mechanism(s) of tumor promotion since TPA also stimulates oxygen radical production in epidermal cells. It is of interest to note that endogenous SOD, which is thought to protect cells from the toxic effects of free radicals (54), is inhibited in mouse epidermis in vivo by TPA treatment (55). (II) While arachidonate metabolism appears to be clearly a part of the promotion process, it is not clear whether it is particular metabolites or the metabolic process itself which generates reactive oxygen species that are ultimately more important. (III) An excess of oxygen radicals has been known for a long time to have numerous effects, including such events as lipid peroxidation, polysaccharide depolymerization, enzyme activation or inactivation, and DNA strand breaks. Since reactive oxygen species or their products have been implicated in gene activation, it seems reasonable to continue to pursue the hypothesis that at least one of the mechanisms by which TPA can alter proliferation/differentiation is through changes in gene regulation that result from reactive oxygen generated in part by the arachidonate cascade.

Table 7. Free Radical Modifier Effects on TPA-induced Chemiluminescence in Mouse Epidermal Cells[a]

Agent	Concentration		% Inhibition TPA Response
Superoxide dismutase	155	units/ml	85
CuDIPS[b]	10	μM	65
Catalase	57.6	units/ml	7
	576	units/ml	10
Mannitol	100	μM	11
Ethoxyquin HCl	.01	μM	25
	0.1	μM	51
	1.0	μM	63
Retinoic acid	10	μM	13
	100	μM	22

[a]10^7 murine epidermal cells were used per 3 ml assay; the cpm at the peak of the responses were used to determine percent inhibition. Experiments were done twice, on different cell isolations.
[b]Copper (II) (3,5-diisopropylsalicylate)2
[c]Retinoic acid analog: ethyl all trans-9-(4-methoxy-2,3,6 trimethylphenyl)-3-7-dimethyl-2,4,6,8 nonatetraenoate.

Table 8. Effect of Inhibitors of Arachidonate Metabolism on TPA-induced Chemiluminescence in Mouse Epidermal Cells.[a]

drug	dose (μM)	% TPA response
Phenidone	1	42
	10	26
	100	0
ETYA[b,d]	1	89
	10	84
	100	9
NDGA[c]	.01	107
	1	90
	10	65
	100	0
Benoxaprofen[d]	33	35
	100	23
	300	2
Indomethacin[d]	.1	111
	1	99
	10	106
	100	84
Flurbiprofen	1	110
	10	93
	100	79

[a] 10^7 murine epidermal cells were used per 3 ml assay; the cpm at the peak of the response was used to determine percent of the TPA response. Experiments were done twice, on different cell isolations.
[b] 5,8,11,14 eicosatetraynoic acid
[c] nordihydroguaiaretic acid
[d] preincubation used

V. CONCLUSION

The above studies indicate that while the prostaglandins or other metabolites of arachidonic acid metabolism that have been tested to date are not tumor promoters themselves, they play an integral part in the underlying mechanism of tumor promotion. The thromboxanes, prostacyclin and leukotrienes have not yet been investigated; information on the contribution of these to tumor promotion will be of value. Tumor promotion studies using the phorbol ester TPA in SENCAR mice have shown that (I) exogenous application of various prostaglandins can modify the tumor yield, the direction being dependent on the particular prostaglandin used, and (II) the lipoxygenase pathway of arachidonic acid metabolism is important if not essential to promotion. An _in vitro_ epidermal model system has been established such that there is good correlation in the response to arachidonate inhibitors between _in vitro_ DNA synthesis and PGE_2 and HPETE production and _in vivo_ DNA synthesis and tumor promotion. Studies on the TPA-induced generation of reactive oxygen species in human PMNs and in mouse epidermal cells have indicated that the same inhibitors can diminish this response considerably, suggesting that the arachidonate cascade may be a source of the free radicals that are thought to play a role in tumor promotion. Additional work is clearly needed to clarify the role(s) of specific arachidonate metabolites or free radical generating metabolic processes in tumor promotion.

REFERENCES

1. Berenblum, I. and Shubik, P. A new, quantitative approach to the study of the stages of chemical carcinogenesis in the mouse skin. Br. J. Cancer 1:373-391, 1947.
2. Slaga, T.J., Fischer, S.M., Nelson, K., and Gleason, G.L. Studies on the mechanism of skin tumor promotion: Evidence for several stages in promotion, Proc. Natl. Acad. Sci. USA 77: 3659-3663, 1980.
3. Slaga, T.J., Fischer, S.M., Weeks, C.E., and Klein-Szanto, A.J.P. In: E. Hodgson, J.R. Bend, R.M. Philpot (eds.), Rev. Biochem. Tox., Elsevier/North Holland, New York, 1981, pp. 231-282.

4. Slaga, T.J. Mechanisms involved in two-stage carcinogenesis in mouse skin. In: Slaga T.J. (ed.), Mechansims of Tumor Promotion. CRC Press Inc., Boca Raton FL, 1984, pp. 1-16.
5. Slaga, T.J., Klein-Szanto, A.J.P., Fischer, S.M., Weeks, C.E., Nelson, K., and Major, S. Studies on the mechanism of action of anti-tumor promoting agents: Their specificity in two-stage promotion. Proc. Natl. Acad. Sci. USA 77:2251-2254, 1980.
6. Slaga, T.J., Fischer, S.M., Viaje, A., Berry, D.L., Bracken, W.M., LeClerc, S., and Miller, D.L. Inhibition of tumor promotion by anti-inflammatory agents: An approach to the biochemical mechanism of promotion. In: T.J. Slaga, A. Sivak and R.K. Boutwell (eds.), Carcinogenesis Vol. 2, Mechanisms of Tumor Promotion and Cocarcinogenesis, Raven Press, New York, 1978, p. 173-195.
7. Slaga, T.J., Fischer, S.M., Weeks, C.E., and Klein-Szanto, A.J.P. Multistage chemical carcinogenesis, In: M. Seije and I.A. Bernstein (eds.), Biochemistry of Normal and Abnormal Epidermal Differentiation, University of Tokyo Press, Tokyo, 1980, pp. 19-218.
8. Janoff, A., Klassen, A., and Troll, W. Local vascular changes induced by the cocarcinogen phorbol myristate acetate. Cancer Res. 30:2568, 1970.
9. Marks, F., Berry, D.L., Bertsch, S., Furstenberger, G., and Richter, H. In: E. Hecker, F. Marks, N. Fusenig and T.J. Slaga, (eds.), Cocarcinogenesis and Biological Effects of Tumor Promoters, Raven Press, N.Y., 1982, pp. 331-246.
10. Schwartz, J.A., Viaje, A., Slaga, T.J., Yuspa, S.H., Hennings, H., and U. Lichti. Fluocinolone acetonide: A potent inhibitor of mouse skin tumor promotion and epidermal DNA synthesis. Chem. Biol. Interact. 17:331, 1977.
11. Hong, S.L. and Levine, L. Inhibition of arachidonic acid release from cells as the biochemical action of anti-inflammatory corticosteroids. Proc. Natl. Acad. Sci. 73:1730, 1976.
12. Gryglewski, R.J. Effects of anti-inflammatory steroids on arachidonate cascade. In: G. Weissman et al. (eds.), Adv. Inflammation Res., Vol. 1, Raven Press; New York, 1979, p. 505.
13. Belman, S. and Troll, W. The inhibition of croton oil-promoted mouse skin tumorigenesis by steroid hormones. Cancer Res. 32: 450-454, 1972.

14. Scribner, J.D. and Slaga, T.J. Multiple effects of dexamethasone on protein synthesis and hyperplasia caused by a tumor promoter. Cancer Res. 33:542-546, 1973.
15. Rohrschneider, L.R. and Boutwell, R.K. The early stimulation of phospholipid metabolism by 12-O-tetradecanoylphorbol-13-acetate and its specificity for tumor promotion. Cancer Res. 33:1945, 1973.
16. Rohrschneider, L.R., O'Brien, T.H., and Boutwell, R.K. The stimulation of phospholipid metabolism in mouse skin following phorbol ester treatment. Biochim. Biophys. Acta. 280:57, 1972.
17. Wertz, P.W. and Mueller, G.C. Inhibition of 12-O-tetradecanoylphorbol-13-acetate-accelerated phospholipid metabolism by 5,8,11, 14-eicosatetraynoic acid. Cancer Res. 40:776, 1980.
18. Glavind, J. and Arffmann, E. The possible carcinogenic properties of altered lipids. Acta Path. Microbiol. Scand. Sect. A 78: 345, 1970.
19. Arffman, E. and Glavind, J. Tumor promoting activity of fatty acid methyl esters in mice. Experientia 27:1765, 1971.
20. Rohrschneider, L.R. and Boutwell, R.K. Phorbol esters, fatty acids and tumor promotion. Nature (New Biology) 243:212, 1973.
21. Furstenberger, G. and Marks, F. Tumor promoter 12-O-tetradecanoylphorbol-13-acetate is not a prostaglandin E-type agonist. Cancer Letters, 6:73-77, 1979.
22. Brune, K., Kalin, H., Schmidt, R., and Hecker, E. Inflammatory, tumor initiating and promoting activities of polycyclic aromatic hydrocarbons and diterpene esters in mouse skin as compared with their prostaglandin releasing potency in vitro. Cancer Letters, 4:333, 1978.
23. Bresnick, E., Meunier, P. and Lamden, M. Epidermal prostaglandins after topical application of a tumor promoter. Cancer Letters 7: 121-125, 1979.
24. Verma, A.K., Ashendel, C.L., Boutwell, R.K. Inhibition by prostaglandin synthesis inhibitors of the induction of epidermal ornithine decarboxylase activity, the accumulation of prostaglandins and tumor promotion caused by 12-O-tetradecanoylphorbol-13-acetate. Cancer Res. 40:308-315, 1980.
25. Boutwell, R.K. Biochemical mechanisms of tumor promotion. In: T.J. Slaga, A. Sivak, and R.K. Boutwell (eds.), Carcinogenesis a Comprehensive Survey, Vol. 2, Raven Press, N.Y., 1978, pp. 49-58.

26. Verma, A.K., Rice, H.M., and Boutwell, R.K. Prostaglandins and skin tumor promotion: Inhibition of tumor promoter-induced ornithine decarboxylase activity in epidermis by inhibitors of prostaglandin synthesis. Biochem. Biophys. Res. Commun. 79:1160-1166, 1977.
27. Nakadate, T., Yamamoto, S., Ishii, M., and Kato, R. Inhibition of 12-O-tetradecanoyl-phorbol-13-acetate-induced epidermal ornithine decarboxylase activity by lipoxygenase inhibitors: possible role of product(s) of lipoxygenase pathway. Carcinogenesis 3: 1411-1414, 1982.
28. Nakadate, T., Yamamoto, S., Ishii, M., and Kato, R. Inhibition of 12-O-tetradecanoylphorbol-13-acetate-induced epidermal ornithine decarboxylase activity by phospholipase A_2 inhibitors and lipoxygenase inhibitor. Cancer Res. 42:2841-2845, 1982.
29. Fischer, S.M., Gleason, G.L., Bohrman, J.S., and Slaga, T.J. Prostaglandin modulation of phorbol ester skin tumor promotion. Carcinogenesis 1:245-248, 1980.
30. Furstenberger, G. and Marks, F. Studies on the role of prostaglandins in the induction of cell proliferation and hyperplasia and in tumor promotion in mouse skin. In: E. Hecker et al., (eds.), Symp. on Cocarcinogenesis and Biological Effects of Tumor Promoters. Raven Press, N.Y., 1982, pp. 325-330.
31. Fischer, S.M., Gleason, G.L., Mills, G.D., and Slaga, T.J. Indomethacin enhancement of TPA tumor promotion in mice. Cancer Letters 10:343,350, 1980.
32. Fischer, S.M. The role of prostaglandins in tumor promotion. In: T.J. Slaga (ed.), Mechanisms of Tumor Promotion, Vol. II Tumor Promotion and Skin Carcinogenesis, CRC Press, Inc., Boca Raton, 1983, pp. 113-126.
33. Fischer, S.M., Mills, G.D., and Slaga, T.J. Inhibition of mouse skin tumor promotion by several inhibitors of arachidonic acid metabolism. Carcinogenesis 3:1243-1245, 1982.
34. Nakadate, T., Yamamoto, S., Iseki, H., Sonoda, S., Takemura, S., Ura, A., Hosoda, Y., and Kato, R. Inhibition of 12-O-tetradecanoylphorbol-13-acetate-induced tumor promotion by nordihydroguaiaretic acid, a lipoxygenase inhibitor and p-bromophenacyl bromide, a phospholipase A_2 inhibitor. Gann 73:841-843, 1982.
35. Kato, R., Nakadate, T., Yamamoto, S., and Sugimura, T. Inhibition of 12-O-tetradecanoylphorbol-13-acetate-induced tumor promotion

and ornithine decarboxylase activity by quercetin: possible involvement of lipoxygenase inhibition. Carcinogenesis 4:13011305, 1983.

36. Hamberg, M. and Samuelsson, B. Prostaglandin endoperoxides. Novel transformations of arachidonic acid in human platlets. Proc. Natl. Acad. Sci. 71:3400, 1974.

37. Hammarstrom, S., Hamberg, M., Samuelsson, B., Duell, E., Stawiski, M., and Vorhees, J.J. Increased concentrations of nonesterified arachidonic acid, 12L-hydroxy 5,8,10,14-eicosatetraenoic acid, prostaglandin E_2 and prostaglandin $F_{2\alpha}$ in epidermis of psoriasis, Proc. Natl. Acad. Sci. USA, 72:5130, 1975.

38. Hamberg, M. and Samuelsson, B. Prostaglandins in human seminal plasma. J. Biol. Chem. 241:257, 1966.

39. Fischer, S.M. and Slaga, T.J. Modulation of prostaglandin synthesis and tumor promotion. In: T.J. Poweles et al. (eds.), Prostaglandins and Cancer First International Conference, Alan R. Liss, Inc., New York, 1982, pp. 255-264.

40. Colburn, N.H., Vorder Bruegge, W.F., Bates, J.R., Gray, R.H., Rossen, J.D., Kelsey, W.H., and Shimada, T. Correlation of anchorage independent growth with tumorigenicity of chemically transformed mouse epidermal cells. Cancer Res. 38:624-634, 1978.

41. Slaga, T.J., Viaje, A., Bracken, W.M., Buty, S.G., Miller, D.R., Fischer, S.M., Richter, C.K., and Dumont, J.N. In vitro transformation of epidermal cells from newborn mice. Cancer Res. 38:2246-2252, 1978.

42. Fischer, S.M., Viaje, A., Harris, K.L., Miller, D.R., Bohrman, J.S. and Slaga, T.J. Improved conditions for murine epidermal cell culture. In Vitro 16:180-188, 1980.

43. Yotti, L.P., Chang, C.C., and Trosko, J.E. Elimination of metacooperation in Chinese hamster cells by a tumor promoter. Science 206, 1989-1091, 1979.

44. Murray, A.W. and Fitzgerald, D.J. Tumor promoters inhibit metabolic cooperation in cocultures of epidermal and 3T3 cells. Biochem. Biophys. Res. Commun. 91:395, 1979.

45. Slaga, T.J., Klein-Szanto, A.J.P., Triplett, L.L. and Yotti, L.P. Skin tumor-promoting activity of benzoyl peroxide, a widely used free radical-generating compound. Science 213:1023-1025, 1981.

46. Fischer, S.M., Mills, G.D., and Slaga, T.J. In: B. Sammuelsson, R. Paoletti, and P. Ramwell (eds.), Adv. Prostaglandin, Thromboxane and Leukotriene Research, Vol. 12, Raven Press, New York, 1983, pp. 309-312.
47. Kensler, T.W., Bush, D.M. and Kozumbo, W.J. Inhibition of tumor promotion by a biomimetic superoxide dismutase. Science 221: 75-77, 1983.
48. Kensler, T.W. and Trush, M.A. Inhibition of oxygen radical metabolism in phorbol ester-activated poly morphonuclear leukocytes by an antitumor promoting copper complex with superoxide dismutase mimetic activity. Biochem. Pharmacol. 32:3485-3487, 1983.
49. Goldstein, B.D., Witz, G., Amoruso M., and Troll, W. Protease inhibitors antagonize the activation of polymorphonuclear leukocyte oxygen consumption. Biochem. Biophys. Res. Commun. 88:854-860, 1979.
50. Goldstein, B.D., Witz, G., Amoruso, M., Stone, D.S., and Troll, W. Stimulation of human polymorphonuclear leukocytes superoxide anion radical production by tumor promoters. Cancer Letters 11:154, 1981.
51. Kensler, T.W. and Trush, M.A. Inhibition of phorbol esterstimulated chemiluminescence in human polymorphonuclear leukocytes by retinoic acid and 5,6 epoxyretinoic acid. Cancer Res. 41:216-222, 1981.
52. Fischer, S.M. and Adams, L.A. Tumor promoter-induced chemiluminescence in mouse epidermal cells is inhibited by several inhibitors of arachidonic acid metabolism. (submitted)
53. Trush, M.A., Wilson, M.E., and Van Dyke, K. The generation of chemiluminescence by phagocytic cells. Methods Enzymol. 57:462-494, 1978.
54. DeChatelet, L.R., McCall, C.E., McPhail, L.C. and Johnston, R.B. Superoxide dismutase activity in leukocytes. J. Chm. Invest. 53: 1197-1201, 1974.
55. Solanki, V., Rana, R.S., and Slaga, T.J. Diminuition of mouse superoxide dismutase and catalase activities by tumor promoters. Carcinogenesis 2:1141-1146, 1981.

Chapter 3

PROSTAGLANDINS, EPIDERMAL HYPERPLASIA AND SKIN TUMOR PROMOTION
Gerhard Furstenberger and Friedrich Marks
German Cancer Research Institute
Heidelberg, Germany

		Page
I.	Introduction	50
II.	The role of prostaglandins in epidermal cell proliferation in vivo	50
	A. Epidermal cell proliferation in normal adult epidermis	50
	B. Induction of epidermal hyperproliferation and hyperplasia in adult mouse skin	52
	C. Epidermal cell proliferation in neonatal mouse skin	54
III.	The role of prostaglandins in epidermal cell proliferation in vitro	55
IV.	The role of prostaglandins in the process of tumor promotion	57
	A. Two-stage tumor promotion in mouse skin	59
	B. The role of prostaglandins in the first stage of promotion	60
	C. The role of prostaglandins in the second stage of promotion	64
V.	A look at the products of lipoxygenase pathways in epidermal cell proliferation and in tumor promotion	66
References		69

S.M. Fischer and T.J. Slaga (eds.), ARACHIDONIC ACID METABOLISM AND TUMOR PROMOTION. Copyright © 1985. Martinus Nijhoff Publishing, Boston. All rights reserved.

3. PROSTAGLANDINS, EPIDERMAL HYPERPLASIA AND SKIN TUMOR PROMOTION
G. FÜRSTENBERGER AND F. MARKS

I. INTRODUCTION

Prostaglandins are a family of hormone like compounds derived from 20-carbon polyunsaturated fatty acids such as arachidonic acid. Besides other oxygenated arachidonic acid derivatives such as hydroxy-, hydroperoxy fatty acids and leukotrienes, prostaglandins have been reported as being produced in many tissues (for review see 1,2). Generally, production in a given tissue is initiated by an exogenous or endogenous stimulus and the prostaglandins are modifiers of tissue response to that stimulus. The short half life of these compounds limits their activity to the site of production. As biologically potent locally acting compounds prostaglandins may play important regulatory roles in individual cell functions, cell interactions and tissue homeostasis.

The scope of this paper is to summarize the results of our studies directed towards the role of prostaglandins in the induction of epidermal hyperproliferation and hyperplasia by exogenous stimuli, and in the process of tumor promotion in mouse skin.

II. THE ROLE OF PROSTAGLANDINS IN EPIDERMAL CELL PROLIFERATION IN VIVO

In the skin, prostaglandins and other arachidonic acid derived compounds have been shown to be involved in inflammatory processes (3,4), epidermal growth with respect to wound repair (5) and in proliferative skin diseases such as psoriasis (6,7). In mouse skin, prostaglandins play a critical role in the induction of epidermal growth processes by a variety of mechanical and chemical stimuli (8).

IIA. Epidermal cell proliferation in normal adult mouse epidermis

Adult mouse epidermis in vivo (strain NMRI) transforms arachidonic acid via the cyclooxygenase pathway to prostaglandins E_2, $F_{2\alpha}$ and D_2 as determined by the isolation of prostaglandins and their identification by radioimmunoassay (8) and HPLC analysis (unpublished results). When topically applied to normal adult mouse skin, none of these prostaglandins influence significantly either epidermal DNA synthesis (Fig. 1) or mitotic activity (8). Even the metabolically more stable 15S-15-methyl-PGE$_2$ or 15-S-15-methyl-PGF$_{2\alpha}$ were completely inactive as mitogens (Fig. 1). These results are clearly in contrast to data published by Lowe and Stoughton (9). The reasons for this discrepancy are not yet clear. Vice versa, normal epidermal DNA synthesis and mitotic activity are not inhibited by doses of the cyclooxygenase inhibitor indomethacin, which decrease epidermal prostaglandin levels (10). This indicates that normal, everyday epidermal cell proliferation does not depend on prostaglandin activities.

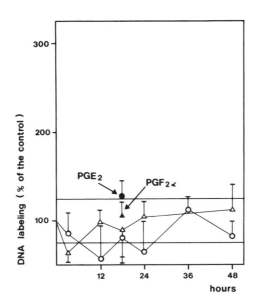

FIGURE 1. Effects of prostaglandins E$_2$, F$_{2\alpha}$, and of 15-S-15-methyl PGE$_2$ (●) and 15-S-15-methyl PFG$_{2\alpha}$ (▲) on epidermal DNA synthesis. Mice were topically treated with 0.1 ml acetone, with 10 µg of either PGE$_2$ and PGF$_{2\alpha}$ in 0.1 ml acetone or with 15 µg 15-S-15-methyl PGE$_2$ and 15 µg 15-S-15-methyl PGF$_{2\alpha}$ in 0.1 ml acetone. The animals were killed at the times indicated. One hour prior to sacrifice, labeled thymidine was i.p. injected (for experimental details, see 11). Each point represents the mean of at least 6 mice ±S.D. The average standard deviation of the controls (52±10; n=40) is indicated by horizontal lines.

Rodent skin contains specific membrane receptors for both E- and F-prostaglandins (12). Since topically applied prostaglandin E_2 (PGE_2) has been shown to result in a weak increase of epidermal cyclic AMP level, the prostaglandin E receptor complex of mouse epidermis may interact with adenylate cyclase (13). Prostaglandin $F_{2\alpha}$ ($PGF_{2\alpha}$) and prostaglandin D_2 (PGD_2) are inactive in this respect. The physiological role of cyclic AMP, and thus the physiological significance of this effect in skin, is not known.

IIB. Induction of epidermal hyperproliferation and hyperplasia in adult mouse skin

Whereas prostaglandins are apparently not involved in normal mouse epidermis cell proliferation, they have been shown to play a crucial role in the induction of epidermal hyperplasia (8,14). Hyperplastic growth is the general response of mouse skin to mechanical or chemical irritation, for example using the irritant tumor promoter 12-O-tetradecanoylphorbol-13-acetate (TPA). Besides strong inflammatory effects, TPA induces epidermal hyperproliferation, including induction of ornithine decarboxylase (ODC) activity (15), which is consistently associated with epidermal hyperplasia (hyperplastic transformation; 16). Two waves of epidermal prostaglandin E synthesis occurring 10 and 90 minutes after treatment are observed as one of the earliest responses to TPA, perhaps corresponding to the stimulation of different epidermal cell populations (11). At later times after stimulation, the level of epidermal prostaglandin E has been found to be increased up to 4 days following a single TPA treatment (17). Experiments with the cyclooxygenase inhibitor indomethacin have shown that early prostaglandin E synthesis is essential for the induction of epidermal ODC activity and hyperplasia (11,14,18). A more detailed analysis has revealed that the first wave of PGE accumulation (10 min.) was essential for hyperproliferation, in that prevention of the first, but not of the second wave of PGE synthesis by indomethacin, results in an inhibition of TPA-induced hyperproliferation (11). PGE_2 and, to a lesser extent, PGD_2 (but not $PGF_{2\alpha}$) are able to release the indomethacin induced inhibition of ODC activity (18) and epidermal hyperproliferation in a specific and highly sensitive manner (11,14). These data are schematically summarized in Fig. 2.

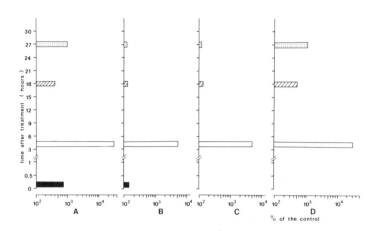

FIGURE 2. Effect of indomethacin on TPA-induced epidermal prostaglandin E content (■), ODC activity (□), DNA labeling (▨) and mitotic activity (▥) and its reversion by PGE_2 but not by $PGF_{2\alpha}$. Mice were locally treated with acetone one hour prior to TPA (10 nmol) [A] or indomethacin (1.1 μmol/100 μl acetone) one hour before application of TPA (10 nmol) [B], TPA (10 nmol) + PGE_2 [D] or TPA (10 nmol) + $PGF_{2\alpha}$ [C] dissolved in 0.1 ml acetone. The mice were killed at the times indicated. For determination of epidermal prostaglandin E content, ODC activity, DNA labeling and mitotic activity, see 11,14. Each value is the mean of at least 6 mice.

Notwithstanding the fact that prostaglandins do not stimulate epidermal cell proliferation in untreated skin, PGE_2 shows a comitogenic effect when applied simultaneously with TPA (19). A similar comitogenic effect has been found to be exerted by $PGF_{2\alpha}$ although this prostaglandin is unable to overcome indomethacin inhibition of TPA-induced hyperplasia. These results indicate that TPA gives rise to both PGE_2- and $PGF_{2\alpha}$-responsive cells in the course of stimulation of epidermal cell proliferation. Both E- and F-type prostaglandins are apparently involved in the hyperproliferative response to TPA treatment, with prostaglandin E being essential for the "trigger"-process, i.e. the transformation of normal epidermis into the hyperplastic state. Besides PGE_2 accumulation, TPA induces two waves of epidermal prostaglandin F synthesis peaking around 2.5 and 4 hours after treatment (8).

The induction of epidermal hyperplastic growth seems to be due to a synergistic action of a given stimulus and PGE_2, since otherwise PGE_2 should not only be able to overcome indomethacin inhibition, but also to induce a hyperplastic response by itself in untreated mouse skin. This is not the case. The mechanism of this synergistic effect remains to be elucidated. There is experimental evidence that the synergistic effect of PGE_2 on the induction of epidermal hyperproliferation is not mediated by cyclic AMP (13).

An involvement of prostaglandins in the induction of hyperplastic growth is not only observed after treatment with the tumor promoter TPA and other irritant skin mitogens such as the ionophore A 23187 (14), but also after mechanical stimulation of adult mouse epidermis. For instance prostaglandin involvement occurs after removal of the horny layer by sandpaper rubbing (20) or full skin wounding (21). Actually, prostaglandin-mediated transformation seems to be the general response of mouse skin to irritation and damage.

Non-irritant and non-damaging stimuli such as treatment with 4-O-methyl-TPA (22) or skin massage (20) induce a type of epidermal hyperproliferation which has been found to be resistant to indomethacin inhibition and not to be accompanied by prostaglandin accumulation. In addition, PGE_2 and $PGF_{2\alpha}$ do not exhibit a comitogenic effect in skin treated with 4-O-methyl-TPA. Interestingly, the prostaglandin-independent hyperproliferation of mouse skin does not result in epidermal hyperplasia (22).

Endogenous prostaglandin synthesis has also been reported to be involved in the induction of DNA synthesis and in mitotic activity of hepatocytes after partial hepatectomy (23). Moreover, indomethacin inhibits estrogen-induced DNA synthesis in rat uterus indicating that the hormonal effect is mediated by prostaglandins. Indeed, estrogen stimulates the uterine production of PGE_2 and $PGF_{2\alpha}$ in rats subject to ovarectomy (24).

IIC. Epidermal cell proliferation in neonatal mouse skin

Neonatal mouse epidermis differs from adult mouse epidermis in that it is resistant to the growth stimulatory effects evoked by TPA or by removal of the horny layer (16,25). Accordingly, TPA does not produce an early elevation of prostaglandin E levels nor induces ODC activity in neonatal mouse epidermis (26 and unpublished results). PGE_2 and $PGF_{2\alpha}$ exhibit neither a mitogenic effect when applied alone, nor a comitogenic effect when given in combination with TPA.

In contrast to superficial damage or chemical irritation, full skin wounding induces a proliferative response in neonatal mouse skin as it does in adult skin (21). Interestingly, the stimulation of cell proliferation in neonatal mouse epidermis is insensitive to indomethacin, whereas in adult skin, it is partially inhibited by the drug (21). This indicates that prostaglandins are not involved in the wound response of neonatal mouse skin. Thus, prostaglandin E-mediated hyperplastic transformation is restricted to adult mouse epidermis. Seen ontogenetically, hyperplastic transformation develops within the first week after birth (26). One explanation for the resistance of neonatal mouse skin to hyperplastic stimulation could be that the prostaglandin-synthesizing enzymatic machinery is not yet fully developed. On the other hand, homogenates of neonatal mouse epidermis transform arachidonic acid via the cyclooxygenase pathway to prostaglandins E_2, $F_{2\alpha}$ and D_2 (unpublished results).

III. THE ROLE OF PROSTAGLANDINS IN EPIDERMAL CELL PROLIFERATION IN VITRO

In cultured epidermal cells, TPA has been shown to induce a similar sequence of biochemical events as in vivo, i.e. stimulation of prostaglandin synthesis, induction of ornithine decarboxylase activity and DNA synthesis. The data obtained with the murine epidermal cell line HEL-30 (27), primary basal epidermal cells of the adult guinea pig ear (28) and primary basal cells of the neonatal mouse (29) are summarized in Fig. 3.

In these three cell types TPA activates phospholipase A_2 which results in a release of arachidonic acid from $[1-^{14}C]$-arachidonic acid prelabeled phospholipids (Fig. 3A). The activation of phospholipase A_2 activity requires extracellular Ca^{2+} (30). Since TPA does not exhibit any measurable effect on Ca^{2+}-cell influx, it may be concluded that the stimulation of phospholipase A_2 is due to a mobilization of cell surface bound Ca^{2+}. Such a mechanism is also indicated by the inhibition of the TPA-induced arachidonic acid release by verapamil or tetracaine (30), i.e. both drugs which are thought to interact with Ca^{2+}-binding sites at cell surfaces (31).

In HEL-30 cells and primary basal keratinocytes of neonatal mouse skin, TPA stimulates the release of arachidonic acid and PGE_2 into the culture medium after a lag phase of 10 minutes, whereas a significant elevation of the $PGF_{2\alpha}$ content was observed only after one hour. Following TPA

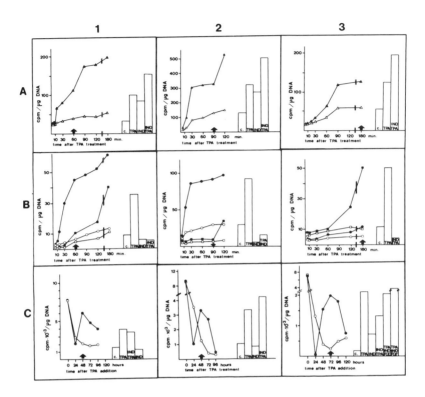

FIGURE 3. Effects of TPA, indomethacin and prostaglandins E_2 and $F_{2\alpha}$ on the accumulation of arachidonic acid (A), prostaglandins E_2 and $F_{2\alpha}$ (B; ●○,■□), and the stimulation of DNA labeling (C) using HEL-30 (1), primary basal epidermal cells of neonatal mouse skin (2) and the adult guinea pig ear (3). For determination of arachidonic acid and prostaglandins, cultures were prelabeled with $[1-^{14}C]$ arachidonic acid. Cells were treated with 10^{-7}M TPA (closed symbols), indomethacin (10^{-4}M), PGE$_2$ (10^{-6}M), PGF$_{2\alpha}$ (10^{-6}M) or acetone or DMSO (open symbols) at zero time and processed at the times indicated for measurement of arachidonic acid and prostaglandin content in the culture medium and DNA labeling. For methodology see ref. 27.

treatment of guinea pig adult primary basal keratinocytes, arachidonic acid and PGF$_{2\alpha}$ are released into the medium after a lag phase of 30 minutes (Fig. 3A,B). All three cultures respond to TPA treatment by an induction of ODC activity. Indomethacin inhibits TPA-induced prostaglandin release and, at the same time, enhances the release of arachidonic acid into the culture medium (Fig. 3A,B). In contrast to the situation in mouse epidermis in vivo (15), TPA-induced ODC-activity is not inhibited by indomethacin treatment of the three cell cultures. The same is true for the

TPA-enhanced DNA labeling in the mouse cells, whereas in the guinea pig cells a partial inhibition of indomethacin was observed (Fig. 3C). This inhibition could be reversed by $PGF_{2\alpha}$ and, to a lesser extent, also by PGE_2.

In epidermal cell cultures, we observe either partial or complete uncoupling of prostaglandin synthesis from ODC induction and epidermal hyperproliferation evoked by TPA. One possible explanation for this difference between the in vivo and in vitro situation is based on the consideration that prostaglandins are amplifiers rather than mediators of the TPA effect. If TPA exerts hormone-like effects or facilitates the interaction of keratinocytes with unknown endogenous factors, it could well be that, in the living animal, such an effect has to be amplified by prostaglandins, whereas in cell culture the condition may be favorable enough not to require amplification. A similar uncoupling of TPA-induced prostaglandin synthesis from ODC induction, as well as DNA synthesis, has been observed in 3T3 cells (32).

In contrast to other published data (33), we could not find PGF_2 and $PGF_{2\alpha}$ stimulatory effects on ODC activity and on DNA labeling using either HEL-30 cells or primary keratinocytes. In primary cultures of neonatal rat liver and in T 51 B rat liver cells, arachidonic acid and prostaglandins E_1 and E_2 were found to stimulate mitotic acitvity (34,35). $PGF_{2\alpha}$ was reported to induce the initiation of DNA synthesis either by itself, as in qiescent 3T6 and Swiss 3T3 cells, or synergistically with growth factors such as EGF (36).

IV. THE ROLE OF PROSTAGLANDINS IN THE PROCESS OF TUMOR PROMOTION

TPA does not only induce skin inflammation and hyperplastic growth of mouse epidermis, but in addition, promotes the development of mouse skin tumors initiated with a subthreshold dose of a carcinogen. This is most commonly 7,12-dimethylbenz(a)anthracene (DMBA; for review see 37). Experiments with inhibitors of arachidonic acid metabolism indicate a crucial role of arachidonic acid metabolites in the process of tumor promotion (8,18,38-40). With respect to the cyclooxygenase pathway, Boutwell's and our group have shown that indomethacin applied prior to TPA inhibits, in a dose-dependent manner, the formation of tumors in CD-1 and NMRI mice (8,18). However, in Sencar mice, indomethacin has been found rather to promote, than to inhibit tumor promotion (41). The argu-

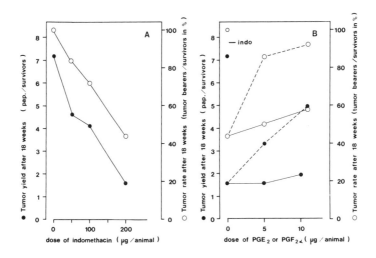

FIGURE 4. Effects of indomethacin (A) and indomethacin plus prostaglandins E_2 and $F_{2\alpha}$ (B) on TPA-induced tumor promotion. For each group, 16 female NMRI mice (7 weeks old) were initiated by topical application of 100 nmol DMBA (dissolved in 0.1 ml acetone) onto the shaved back-skin. One week after DMBA application, the mice were topically treated with acetone or various doses of indomethacin dissolved in 0.1 ml acetone 60 minutes prior to TPA treatment (5 nmol/100 µl acetone). The administration of various doses of PGE_2 (——) and $PGF_{2\alpha}$ (- - -) (dissolved in 0.1 ml acetone) was topically applied simultaneously with TPA (5 nmol/0.1 ml acetone). This treatment occurred twice weekly. The tumor promoting activity is given as tumor yield (number of papillomas/survivor) and tumor rate (number of tumor bearers/survivors). At the end of the experiment 94% or more of the animals were alive.

ment that the inhibitory action of indomethacin on CD-1 and NMRI mice could be due to cytotoxicity of the drug (41) is ruled out by the fact that the inhibition of tumor formation by indomethacin can be specifically reversed by low doses of $PGF_{2\alpha}$, as depicted in Fig. 4. A better explanation for the discrepancies observed would be an inborn fundamental difference between Sencar mice and other mouse strains (42) in the main route of arachidonic acid metabolism. The major products of arachidonic acid metabolism obtained with or without TPA in Sencar mice epidermal cell cultures appear to be products of the lipoxygenase pathways (43), whereas in cultures of NMRI mouse epidermis products of the cyclooxygenase pathway predominate (see Fig. 3B and unpublished results). Moreover, the production of lipoxygenase pathway metabolites of arachidonic acid such as 12-hydroperoxyeicosatetraenoic acid has been reported to be increased upon indomethacin treatment of Sencar mouse keratinocytes (43). There is indeed increasing evidence that products of the arachidonic acid-lipoxygenase pathways may be involved in the process critical for tumor promotion (38,40; see below). Although prostaglandins E_2 and F_2 are inactive as mitogens and tumor promoters in NMRI mouse skin (Table 1), PGE_2, when applied simultaneously with TPA, increases promoting activity, whereas $PGF_{2\alpha}$ has no such effect (Table 1). In Sencar mice, both PGE_2 and $PGF_{2\alpha}$ were found to enhance, and PGE_1 to inhibit, the promoting ability of TPA (44). Cocarcinogenic activities of PGE_2 and $PGF_{2\alpha}$ have been found in methylcholanthrene-treated Swiss albino mice (45).

Table 1. Effects of prostaglandins E_2 and $F_{2\alpha}$ on TPA-induced tumor promotion.

Treatment	Tumor formation[a]			
	After 15 weeks		After 18 weeks	
	Rate[b]	Yield[c]	Rate[b]	Yield[c]
TPA (3.08 µg)	100	9.0	100	8.6
TPA (1.54 µg)	60	3.1	67	3.8
TPA (1.54 µg) + PGE_2 (10 µg)	60	5.3	73	6.4
TPA (1.54 µg) + $PGF_{2\alpha}$ (10 µg)	60	3.4	67	4.0
PGE_2 (10 µg)	0	0	0	0
$PGF_{2\alpha}$ (10 µg)	0	0	0	0

[a]For each group, 15 female NMRI mice (7 weeks old) were initiated by topical application of 100 nmol DMBA (dissolved in 0.1 ml acetone) onto the shaved back skin. Treatment with TPA and/or prostaglandins E_2 and $F_{2\alpha}$ was started one week later. The compounds were dissolved in 0.1 ml acetone and applied twice weekly. All tumor promotion experiments have been performed at least twice, yielding similar results.
[b]number of tumor bearers/survivors (94% or more survivors in all groups).
[c]number of papillomas/survivor

IVA. Two-stage tumor promotion in mouse skin

Investigations into the role of prostaglandins in the process of tumor promotion are impaired by the multiplicity of biological effects evoked by phorbol ester tumor promoters, and while tumor promotion is a long term process. A more detailed analysis can now be achieved since - confirming an experimental protocol originally reported by Boutwell (46) - we and others (42,47) have shown that the process of tumor promotion occurs as, at least, two separate steps. Extended studies on the relationship between the chemical structures and the biological activities of phorbol ester tumor promoters have shown that the tumor promoting potency of TPA can be almost completely abolished when conjugated double bonds are introduced into the long chain fatty acids, or when the tetradecanoic acid residue is replaced by retinoic acid (47,48). The ability of the phorbol esters to act as irritant skin mitogens is not impaired by this structural modification. Indeed, both TPA and 12-O-retinoylphorbol-13-acetate (RPA) induce irritation and hyperplasia to a similar degree and along the same prostaglandin E-dependent pathway (47). In the two-stage approach of skin tumor promotion, RPA acts as an incomplete promoter in that the sequential treatment of initiated mouse skin with a few TPA applications, which by themselves are insufficient for promotion (stage I), followed by twice weekly RPA applications over a period of several weeks (stage II), yields tumor responses comparable to those obtained by prolonged TPA treatment (47). Using this experimental protocol, TPA treatment can be even restricted to only one application, thus achieving for the first time the possibility to distinguish those effects of TPA which are critical or "unique" for promotion from those which reflect the pleiotypic activities of a promoter.

IVB. The role of prostaglandins in the first stage of promotion

The discovery of incomplete promoters and the resulting possibility of subdividing the process of promotion into two stages clearly indicates that skin tumor promotion must involve, in addition to an induction of hyperplastic growth and inflammatory responses, additional effects which are exhibited upon a single promoter treatment during the first stage of promotion. Epidermal hyperplasia may have only a "helper function", in that it is necessary for the expression of these additional effects which are critical and obligatory for promotion. This conclusion is supported by the following experiments: Hydroxyurea applied 18 hours after a single TPA treatment (stage I) leads to an almost complete inhibition of both epidermal DNA synthesis and tumor promotion when followed by repetitive RPA treatments (stage II). This inhibition by hydroxyurea is not due to cytotoxic effects of the drug (49). In another approach, transmaternally initiated neonatal mouse skin has been found to be resistant to the tumor promoting effect of TPA (stage I) as long as it does not respond to TPA by induction of epidermal hyperplasia (26). The sensitivity of neonatal mouse skin to the stimulation of both hyperplasia and tumor promotion shows a similar time course of ontogenetic development (26).

A unique and unexpected feature of the first TPA-dependent stage of promotion is its irreversibility (50): no significant decrease of tumor response was observed when the time interval between stage 1 and 2 is increased up to 2 months or even half a year (unpublished), whereas symptoms of irritation and epidermal hyperproliferation cease within 2 weeks.

Since the development of epidermal hyperplasia is a necessary component of the first stage and since the hyperplastic response has been shown to depend on prostaglandin E accumulation in epidermis (see above), inhibition by cyclooxygenase inhibitors of this stage of promotion is to be expected. When applied one hour prior to TPA, indomethacin indeed inhibits not only the induction of epidermal hyperplasia (10,14), but also inhibits tumor development in the course of the two-stage promotion experiment (Table 2). However, PGE_2, which releases the indomethacin inhibition of hyperproliferative response upon its simultaneous application with TPA (10,14), does not reverse the inhibition of promotion. In addition, $PGF_{2\alpha}$ neither overcomes this indomethacin induced inhibition of hyperplastic growth (10,14) nor that of promotion (Table 2).

A more detailed analysis of this problem has revealed a distinct difference between hyperplastic responses and tumor promotion as far as the prostaglandins are concerned. Whereas for an inhibition of epidermal hyperplasia indomethacin has to be applied prior to TPA treatment, the maximal inhibitory effect on the first stage of promotion is observed when the cyclooxygenase inhibitor is applied 3 hours after TPA (Fig. 5).

Thus, contrary to the hyperproliferative response, for which an almost immediate increase of prostaglandin E synthesis is obligatory, promotion seems to depend on a more delayed indomethacin-sensitive event. This event could be the accumulation of $PGF_{2\alpha}$ observed 3.5 hours after TPA application which is - together with promotion - inhibited by indomethacin applied 3 hours after TPA (Fig. 6). Indeed, indomethacin treatment at later time points, i.e. 6 to 9 hours after TPA, yields only a marginal inhibitory effect on tumor promotion. A critical role of $PGF_{2\alpha}$ is

Table 2. Effect of indomethacin and prostaglandins E_2 and F_2 on the first stage-promoting activity of TPA.

| Treatment | | Tumor formation | | | |
| First (no. of appl.; dose/animal) | Second (no. of appl.; dose/animal) | After 15 weeks | | After 18 weeks | |
		Rate[b]	Yield[c]	Rate[b]	Yield[c]
acetone (2x;100μl) 60 min later TPA (2x;10nmol)	RPA (34x;10nmol)	69	4.8	69	5.3
indomethacin (2x;550nmol) 60 min later TPA (2x;10nmol)	RPA (34x;10nmol)	31	1.9	31	2.5
indomethacin (2x;550nmol) 60 min later TPA (2x; 10nmol) + PGE_2 (2x; 28 nmol)	RPA (34x,10nmol)	19	1.7	33	2.6
indomethacin (2x;550nmol) 60 min later TPA (2x; 10nmol) + $PGF_{2\alpha}$ (2x; 28 nmol)	RPA (34x;10nmol)	38	1.6	38	2.3

[a] For each group 16 female NMRI mice (7 weeks old) were initiated by topical application of 100 nmol DMBA (dissolved in 0.1 ml acetone) onto the shaved back skin. Treatment with indomethacin, prostaglandins E_2 and $F_{2\alpha}$ and TPA was started one week later. The compounds were dissolved in 0.1 ml acetone and applied twice weekly. One week later, treatment with RPA was started. All tumor promotion experiments have been performed at least twice, yielding similar results.

[b] number of tumor bearers/survivors (94% or more survivors in all groups)

[c] number of papillomas/survivor

further indicated by the observation that $PGF_{2\alpha}$ is able to prevent the inhibition of the first stage of promotion, but only when applied simultaneously with indomethacin, i.e. 3 hours after TPA application (Fig. 5). Moreover, the second, perhaps critical wave of prostaglandin F synthesis was not found after application of the incomplete promoter RPA (Fig. 6). These results indicate that the second wave of prostaglandin F synthesis observed 3.5 hours after TPA is critically involved in the first stage of promotion and may turn out to provide the first biochemical parameter for stage 1 of promotion. From these observations one could perhaps conclude that $PGF_{2\alpha}$ applied 3.5 hours after RPA application would restore the tumor promoting efficacy of this incomplete promoter. However, this was not found to be the case (unpublished results).

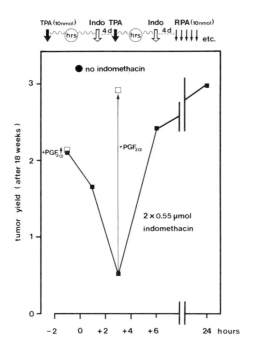

FIGURE 5. Effect on first-stage promotion activity by varying the time interval between indomethacin or indomethacin and PGF$_{2\alpha}$ treatment and TPA application. For each group 16 female NMRI mice (7 weeks old) were initiated by topical application of 100 nmol DMBA (dissolved in 0.1 ml acetone) onto the shaved back skin. One week later, first stage treatment was performed as schematically depicted. Mice were treated with TPA (10 nmol), indomethacin (0.55 μmol) and PGF$_{2\alpha}$ (60 nmol), dissolved in 0.1 ml acetone, followed by repetitive treatments with RPA (10 nmol/0.1 ml acetone). The tumor response is measured as tumor yield, i.e. papillomas/survivors after 18 weeks of promotion. The negative sign refers to indomethacin or indomethacin and PGF$_{2\alpha}$ application before TPA treatment, the positive sign after TPA treatment.

Like TPA, skin wounding exerts a moderate activity as first stage promoter, whereas the ionophore A 23187 and the phorbol ester 4-0-methyl TPA do not reveal significant first stage-promoting potencies (51). The latter most probably due to its inability to induce prostaglandin E-dependent epidermal hyperplasia. In Sencar mice, however, both A23187 and 4-0-methyl TPA have been found to be first stage promoters (52). The reason for these discrepancies is unknown.

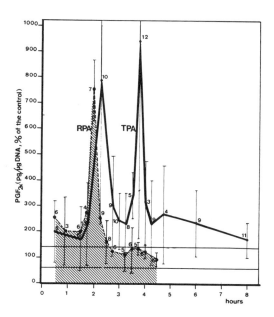

FIGURE 6. Levels of prostaglandin F in mouse epidermis in vivo after topical application of TPA or RPA. Mice were treated with 0.1 ml acetone, 10 nmol TPA or 10 nmol RPA dissolved in 0.1 ml acetone. The mice were killed at the times indicated. The prostaglandins were extracted from frozen epidermis, isolated by thin-layer chromatography and determined by radioimmunoassay (11). Each experimental point represents the mean S.D. for 3-12 mice. The number of animals (experiments) is given beside each point. The horizontal lines represent the average ±S.D. of the control values as obtained with 45 mice (4.3 ± 1.7 PGF/µg DNA = 100%).

IVC. The role of prostaglandins in the second stage of promotion

Stage II of tumor promotion can be performed by one or two applications per week of agents such as RPA or mezerein (42), which are able to induce a prostaglandin-dependent hyperplastic transformation of epidermis (see above). Extending the time interval between RPA applications up to 4 weeks results in a dramatic decrease of tumor development, indicating that the effects of the second stage promoter are reversible in nature and that generation of a sustained epidermal hyperplasia is a necessary and perhaps sufficient condition of stage II promotion. Mitogenic agents which are not capable of inducing epidermal hyperplasia (such as 4-O-methyl TPA), or which do not produce a stationary hyperplasia after repetitive treatments (such as the ionophore A 23187), are ineffective as second stage promoters (51).

Since epidermal hyperplasia is a prostaglandin-dependent process, it is conceivable that indomethacin acts as an inhibitor of the second stage when given in combination with repetitive RPA treatments (Table 3). Considering the fact that the primary induction of epidermal hyperplasia depends specifically on PGE_2 synthesis (11), it was surprising to find that only $PGF_{2\alpha}$, but not PGE_2, applied simultaneously with RPA was able to release the indomethacin inhibition of the second stage of promotion (Table 3).

Table 3. Effects of indomethacin and prostaglandins E_2 and $F_{2\alpha}$ on the second stage of promotion

Treatment		Tumor formation[a]			
First (no.of appl; dose/animal)	Second (no.of appl.; dose/animal)	After 15 weeks		After 18 weeks	
		Rate[b]	Yield[c]	Rate[b]	Yield[c]
TPA (2x; 20 nmol)	RPA (34; 10 nmol)	50	2.6	66	3.1
TPA 2x; 20 nmol)	indomethacin (34x; 550nmol) 60 min later RPA (34x; 10 nmol)	27	1.1	33	1.2
TPA (2x; 20 nmol)	indomethacin (34x; 550nmol) 60 min later $RPA+PGE_2$ (34x; 10nmol + 28 nmol)	27	1.3	27	1.2
TPA (2x; 20 nmol)	indomethacin (34x; 550nmol) 60 min later $RPA+PGF_{2\alpha}$ (34x; 10 nmol + 28 nmol)	60	2.0	66	2.6

[a] For each group 16 female NMRI mice (7 weeks old) were initiated by topical application of 100 nmol DMBA (dissolved in 0.1 ml acetone) onto the shaved back skin. Treatments with TPA were started one week later. Twice-weekly treatments with RPA, indomethacin and prostaglandins E_2 and $F_{2\alpha}$ were started one week after TPA. The compounds were dissolved in 0.1 ml acetone. All tumor promotion experiments have been performed at least twice, yielding similar results.

[b] number of tumor bearers/survivors (94% or more survivors in all groups)

[c] number of papillomas/survivor

One possible interpretation of these results would be that PGE_2 is only necessary to trigger the tissue from the normal, more "quiescent" state into the hyperproliferative state, and that the maintenance of epidermal hyperplasia is then mediated by $PGF_{2\alpha}$. Contrary to the situation during the first stage of promotion, where $PGF_{2\alpha}$ has to be applied several hours after TPA, indomethacin inhibition of the second stage can be reversed when $PGF_{2\alpha}$ is administered simultaneously with RPA. These results and the fact that maximal inhibition of the second stage by indomethacin is obtained by applying indomethacin prior to RPA treatment, may be taken as an indication that $PGF_{2\alpha}$ has different effects on both stages of promotion. Unfortunately, a detailed kinetic analysis of these effects is not possible, since the situation may change profoundly due to repetitive phorbol ester treatments during the second stage.

Taken together, our results indicate that both prostaglandins E_2 and $F_{2\alpha}$ are involved in both stages of tumor promotion, since they are associated with the stimulation of epidermal hyperplasia, which is an integral part of both stages. In addition, $PGF_{2\alpha}$, appearing as one distinct biosynthetic wave at 3.5 hours after TPA treatment, seems to be involved in a step during stage I which is not directly related to hyperplasia but critical for promotion. The distinct prostaglandin peaks observed after phorbol ester application are rather difficult to explain, especially if one considers the fact that TPA stimulates phospholipase A_2 and the arachidonic acid cascade in keratinocytes and in other cell types almost immediately (27,53). One possible explanation would be that different epidermal cell populations are subsequently triggered to proliferate in some kind of a cascade response, and that the delayed activation of one of the prostaglandin F responsive cell populations may provide a critical step in the first stage of promotion.

V. A LOOK AT PRODUCTS OF LIPOXYGENASE PATHWAYS IN EPIDERMAL CELL PROLIFERATION AND IN TUMOR PROMOTION

Besides prostaglandins, products of the lipoxygenase pathways of arachidonic acid are reported to be involved in the mediation of skin inflammation (4,7) and in hyperplastic skin diseases, especially in psoriasis (6,7). Since psoriatic skin is in many morphological and biochemical aspects similar to phorbol ester-treated mouse skin, it is to be expected that products of the lipoxygenase pathways could be involved in TPA-induced epidermal hyperplasia and/or in tumor promotion.

In CD-1 mice, TPA-induced ODC activity, which is generally thought to be a parameter of at least stage II promotion, i.e. epidermal hyperplasia (52,54), has been found to be inhibited by lipoxygenase inhibitors such as nordihydroguaiaretic acid, 3-amino-1-m-(trifluoromethyl)-phenyl-2-pyrazoline and quercetin (39,55). In contrast to the inhibitory effect of indomethacin on ODC induction, inhibition by these agents could not be overcome by prostaglandins, suggesting that both cyclooxygenase and lipoxygenase products might be involved in the induction process (55). Lipoxygenase inhibitors (nordihydroguaiaretic acid and quercetin), as well as inhibitors of both cyclooxygenase and lipoxygenase pathways such as phenidone and 5.8.11.14-eicosatetraynoic acid (ETYA), have been shown to inhibit tumor promotion in CD-1 (38,39) as well as in Sencar mice (40).

In two-stage promotion experiments using NMRI mice, we have found that ETYA in low doses inhibits tumor promotion in a dose-dependent manner (Fig. 7). At the same time such doses inhibit TPA-induced epidermal ODC activity, but not inhibit epidermal DNA labeling evoked by TPA (Fig. 7). Since the latter response is clearly prostaglandin E-mediated, the ETYA doses employed were obviously not able to inhibit the cyclooxygenase pathway of epidermal arachidonic acid metabolism. Therefore, these results indicate that products of the lipoxygenase pathways are involved in processes which are critical for tumor promotion and ODC induction, but not for the hyperproliferative activity induced by TPA. Studies using epidermal homogenates indicate the existence of a dose range where ETYA already inhibits the accumulation of lipoxygenase products such as 12-hydroxy-eicosa-4.8.10.14-tetraenoic acid, but not of cyclooxygenase products such as PGE_2 (unpublished results). Only in higher doses ETYA exhibits both pathways of arachidonic acid metabolism and, in addition, inhibits epidermal hyperproliferation in vivo (14). Accordingly, TPA-induced epidermal ODC activity, which is mediated by both cyclooxygenase and lipoygenase products (56), is at least partially inhibited by ETYA, supporting results obtained with other lipoxygenase inhibitors (39,40,55).

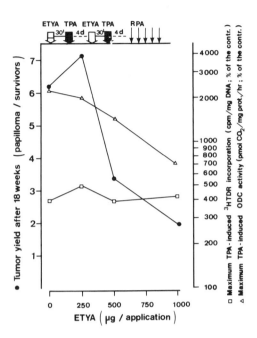

FIGURE 7. Effects of varying the dose of ETYA on TPA-stimulated epidermal ODC activity and DNA labeling, and on the first stage promoting activity exhibited by TPA. Mice were locally treated with acetone (0.1 ml) or varying doses of ETYA, dissolved in 0.1 ml acetone, 30 minutes prior to TPA (10 nmol/0.1 ml acetone). The mice were killed at the time of maximal TPA-induced ODC activity (5 hours post treatment) or maximal TPA-induced epidermal DNA labeling (18 hours post treatment). For the determination of ODC activity and DNA labeling, see ref. 11 and 14. Tumor promotion experiments have been performed using 16 (7 week old) female NMRI mice per group, which were initiated topically by a single dose of 100 nmol DMBA (dissolved in 0.1 ml acetone). One week later, two treatments with varying doses of ETYA/0.1 ml acetone were topically applied 30 minutes prior to TPA (20 nmol/0.1 ml acetone), followed twice weekly by repetitive applications of RPA (10 nmol/0.1 ml acetone). The tumor response is measured as tumor yield after 18 weeks of treatment.

These results indicate that the metabolism of arachidonic acid in mouse epidermis might be involved in the process of tumor promotion via the lipoxygenase, as well as the cyclooxygenase pathway, the latter being predominantly associated with integral, tumor promotive growth processes. However, the precise role of the different metabolites in the process of tumor promotion is still far from being understood.

REFERENCES

1. Samuelsson B, Goldyne M, Granström E, Hamberg M, Hammarström S, Malmsten C: Prostaglandins and thromboxanes. Ann Rev Biochem (47): 997-1029, 1978.
2. Samuelsson B: Prostaglandins, thromboxanes and leukotrienes: biochemical pathways. In: Powles TJ, Bockman RS, Honn KV, Ramwell P (eds) Prostaglandins and Cancer: First International Conference (Prostaglandins and Lipids, Vol 2), Alan R Liss, New York, 1982, pp 1-19.
3. Bonta JL, Parnham MJ: Prostaglandins and chronic inflammation. Biochem Pharmacol (27): 1611-1623, 1978.
4. Samuelsson B: Leukotrienes: A new class of mediators of immediate hypersensitivity reactions and inflammation. In: Samuelsson B, Paoletti R, Ramwell PW (eds) Adv Prostaglandin, Thromboxane and Leukotriene Res Vol 11, Raven Press, New York, 1983, pp 1-13.
5. Anggard E, Jonsson CE: Formation of prostaglandins in the skin following a burn injury. In: Ramwell PW, Pharris BB (eds) Prostaglandins in cellular biology, Vol 1, Plenum Press, New York, 1972, pp 269-291.
6. Hammarström S, Hamberg M, Samuelsson B, Duell M, Stawiski M, Vorhees JJ: Increased concentrations of nonesterified arachidonic acid, 12-L-hydroxy-5,8,10,14-eicosatetraenoic acid, PGE_2 and PGF_2 in epidermis of psoriasis. Proc Natl Acad Sci USA (75): 5130-5134, 1975.
7. Vorhees JJ: Leukotrienes and other lipoxygenase products in the pathogenesis and therapy of psoriasis and other dermatoses. Arch Dermatol (119): 541-547, 1983.
8. Fürstenberger G, Gross M, Marks F: On the role of prostaglandins in the induction of epidermal proliferation, hyperplasia and tumor promotion in mouse skin. In: Powles TJ, Bockman RS, Honn KV, Ramwell P (eds) Prostaglandins and Cancer: First international conference (Prostaglandins and related lipids, Vol 2), Alan R Liss, New York, 1982, pp 239-254.
9. Lowe NJ, Stoughton RB: Effects of topical prostaglandin E_2 analogue on normal hairless mouse epidermal DNA synthesis. J Invest Dermatol (68): 134-137, 1977.
10. Fürstenberger G, Marks F: Indomethacin inhibition of cell proliferation induced by the phorbol ester TPA is reversed by prostaglandin E_2 in mouse epidermis in vivo. Biochem Biophys Res Commun (84): 1103-1108, 1978.
11. Fürstenberger G, Marks F: Early prostaglandin E synthesis is an obligatory event in the induction of cell proliferation in mouse epidermis in vivo by the phorbol ester TPA. Biochem Biophys Res Commun (92): 749-756, 1980.
12. Lord JT, Ziboh VA, Warren SK: Specific binding of prostaglandin E_2 and F_2 by membrane preparations from rat skin. Endocrinology (102): 1300-1309, 1978.
13. Marks F: Prostaglandins, cyclic nucleotides and the effect of phorbol ester tumor promoters on mouse skin in vivo. Carcinogenesis (4): 1465-1470, 1983.
14. Marks F, Fürstenberger G, Kownatzki E: Prostaglandin E-mediated mitogenic stimulation of mouse epidermis in vivo by divalent cation ionophore A 23187 and by tumor promoter 12-O-tetradecanoylphorbol-13-acetate. Cancer Res (41): 696-702, 1981.

15. O'Brien TG: The induction of ornithine decarboxylase as an early, possibly obligatory event in mouse skin carcinogenesis. Cancer Res (36): 2644-2653, 1976.
16. Marks F, Bertsch S, Grimm W, Schweizer J: Hyperplastic transformation and tumor promotion in mouse epidermis: possible consequences of disturbance of endogenous mechanisms controlling proliferation and differentiation. IN: Slaga TJ, Sivak A, Boutwell RK (eds) Carcinogenesis, Vol 2, Raven Press, New York, 1978, pp 97-116.
17. Ashendel CL, Boutwell RK: Prostaglandin E and F levels in mouse epidermis are increased by tumor promoting phorbol esters. Biochem Biophys Res Commun (90): 623-627, 1979.
18. Verma AK, Ashendel CL, Boutwell RK: Inhibition by prostaglandin synthesis inhibitors of the induction of epidermal ornithine decarboxylase activity, the accumulation of prostaglandins and tumor promotion caused by TPA. Cancer Res (40): 308-315, 1980.
19. Fürstenberger G, Gross M, Marks F: Involvement of prostaglandins in the process of skin tumor promotion. In: Proceedings of the international meeting on icosanoids and cancer, Raven Press, New York, 1984, in press.
20. Fürstenberger G, deBravo M, Bertsch S, Marks F: The effect of indomethacin on cell proliferation induced by chemical and mechanical means in mouse epidermis in vivo. Res Commun Chem Pathol Pharmacol (24): 533-541, 1979.
21. Bertsch S, Marks F: A comparative study on wound-healing in neonatal and adult mouse epidermis in vivo. Cell Tissue Kinet (15): 81-87, 1982.
22. Fürstenberger G, Richter H, Argyris TS, Marks F: Effect of the phorbol ester 4-O-methyl TPA on mouse skin in vivo: evidence for its uselessness as a negative control compound in studies on the biological effects of phorbol ester tumor promoters. Cancer Res (42): 342-348, 1982.
23. Rixon RH, Whitfield JF: An early mitosis-determining event in regenerating rat liver and its possible mediation by prostaglandins or thromboxane. J Cell Physiol (113): 281-288, 1982.
24. Stewart PJ, Zalondek CJ, Murphy Inman M, Webster RA: Effects of dexamethasone and indomethacin on estrogen-induced uterine growth. Life Sci (33): 2349-2356, 1983.
25. Bertsch S, Marks F: Removal of the horny layer does not stimulate cell proliferation in neonatal mouse epidermis. Cell Tissue Kinet (11): 651-658, 1978.
26. Marks F, Fürstenberger G: Tumor promotion in skin and the mechanisms involved. In: Davis H (ed), IARC Scientific Publication, Lyon, 1984, in press.
27. Fürstenberger G, Richter H, Fusenig NE, Marks F: Arachidonic acid and prostaglandin E release and enhanced cell proliferation induced by the phorbol ester TPA in a murine epidermal cell line. Cancer Lett (11): 191-198, 1980.
28. Delescluse C, Fürstenberger G, Marks F, Prunieras M: Effects of phorbol esters on basal epidermal cells from adult guinea pig ear skin. Cancer Res (42): 1975-1979, 1982.
29. Fürstenberger G, Gross M, Schweizer J, Vogt I, Marks F: Biochemical and functional characterization of subpopulations of neonatal mouse keratinocytes. Europ J Cancer Clin Oncol (19): 1284, 1983.

30. Ganss M, Seemann D, Fürstenberger G, Marks F: Calcium-dependent release of arachidonic acid from a murine epidermal cell line induced by the tumor promoter TPA or ionophore A 23187. FEBS Letters (142): 54-58, 1982.
31. Low PS, Lloyd DH, Stein ThM, Rogers JA: Calcium displacement by local anesthetics. J Biol Chem (254): 4119-4125, 1979.
32. Lanz R, Brune K: Dissociation of tumor promoter-induced effects on prostaglandin release, polyamine synthesis and cell proliferation of 3T3 cells. Biochem J (194): 975-982, 1981.
33. Bem JL, Greaves MW: Prostaglandin E_1 effects on epidermal cell growth. Arch Dermatol Res (251): 35-41, 1974.
34. Andreis PG, Whitfield JF, Armato U: Stimulation of DNA synthesis and mitosis of hepatocytes in primary cultures of neonatal rat liver by arachidonic acid and prostaglandins. Exp Cell Res (134): 265-272, 1982.
35. Boynton AL, Whitfield JF: Possible involvement of arachidonic acid in the initiation of DNA synthesis by rat liver cells. Exp Cell Res (129): 474-478, 1980.
36. Jimenez de Asua L, Otto AM, Ulrich MO, Martin-Perez J, Thomas G: The regulation of DNA replication in animal cells by PGF_2. In: Powles TJ, Bockman RS, Honn KV, Ramwell P (eds) Prostaglandins and Cancer: First International Conference (Prostaglandins and related lipids, vol 2) Alan R Liss, New York, 1982, pp 309-331.
37. Hecker E, Fusenig NE, Kunz W, Marks, F, Thielmann HW: Cocarcinogenesis and biological effects of tumor promoters. Carcinogenesis, a comprehensive survey, vol 7, Raven Press, New York, 1982.
38. Nakadate T, Yamamoto S, Iseki H, Sonoda S, Takemura S, Ura A, Hosoda Y, Kato R: Inhibition of TPA-induced tumor promotion by nordihydroguaiaretic acid, a lipoxygenase inhibitor and p-bromophenacylbromide, a phospholipase A_2 inhibitor. Gann (73): 841-843, 1982.
39. Kato R, Nakadate T, Yamamoto S, Sugimura T: Inhibition of TPA-induced tumor promotion and ornithine decarboxylase activity by quercetin: possible involvement of lipoxygenase inhibition. Carcinogenesis (4): 1301-1305, 1983.
40. Fischer SM, Mills GD, Slaga TJ: Inhibition of mouse skin tumor promotion by several inhibitors of arachidonic acid metabolism. Carcinogenesis (3): 1243-1245, 1982.
41. Fischer SM, Gleason GL, Mills GD, Slaga TJ: Indomethacin enhancement of TPA tumor promotion in mice. Cancer Lett (10): 343-350, 1980.
42. Slaga TJ, Fischer SM, Weeks CE, Klein-Szanto AJP, Reiners J: Studies on the mechanisms involved in multistage carcinogenesis in mouse skin. J Cell Biochem (18): 99-119, 1982.
43. Fischer SM, Slaga TJ: Modulation of prostaglandin synthesis and tumor promotion. In: Powles TJ, Bockman RS, Honn KV, Ramwell P (eds) Prostaglandins and Cancer: First international conference (Prostaglandins and related lipids, vol 2), Alan R Liss, New York, 1982, pp 255-264.
44. Fischer SM, Gleason GL, Hardin LG, Bohrman JS, Slaga TJ: Prostaglandin modulation of phorbol ester skin tumor promotion. Carcinogenesis (1): 245-248, 1980.
45. Lupulescu A: Enhancement of carcinogenesis by prostaglandins. J Natl Canc Inst (61): 97-106, 1978.

46. Boutwell RK: Some biological effects of skin carcinogenesis. In: Homburger F (ed) Progress in experimental tumor research, Vol 4, Karger, Basel, 1964, pp. 207-247.
47. Fürstenberger G, Berry DL, Sorg B, Marks F: Skin tumor promotion by phorbol esters is a two-stage process. Proc Natl Acad Sci USA (78): 7722-7726, 1981.
48. Marks F, Bertsch S, Fürstenberger G: Ornithine decarboxylase activity, cell proliferation, and tumor promotion in mouse epidermis in vivo. Cancer Res (39): 4183-4188, 1979.
49. Kinzel V, Loehrke H, Goerttler K, Fürstenberger G, Marks F: Suppression of the first stage of 12-O-tetradecanoylphorbol-13-acetate-effected tumor promotion in mouse skin by non-toxic inhibition of DNA synthesis, submitted for publication.
50. Fürstenberger G, Sorg B, Marks F: Tumor promotion by phorbol esters in skin: evidence for a memory effect. Science (220): 89-91, 1983.
51. Fürstenberger G, Marks F: Growth stimulation and tumor promotion in skin. J Invest Dermatol (81): 157s-161s, 1983.
52. Slaga TJ, Klein-Szanto AJP: Initiation-promotion versus complete skin carcinogenesis in mice: importance of dark basal keratinocytes (stem cells). Cancer Investigation (1): 425-436, 1983.
53. Levine L: Stimulation of cellular prostaglandin production by phorbol esters and growth factors and inhibition by cancer chemopreventive agents. In: Powles TJ, Bockman RS, Honn KV, Ramwell P (eds) Prostaglandins and Cancer: First international conference (Prostaglandins and related lipids, Vol 2), Alan R Liss, New York, 1983, pp 189-204.
54. Marks F, Berry DL, Bertsch S, Fürstenberger G, Richter H: On the relationship between epidermal hyperproliferation and skin tumor promotion. In: Hecker E, Fusenig NE, Kunz W, Marks F, Thielmann HW (eds) Cocarcinogenesis and biological effects of tumor promoters (Carcinogenesis, a comprehensive survey, Vol 7), Ravem Press, New York, 1982, pp 331-346.
55. Nakadate T, Yamamoto S, Ishii M, Kato R: Inhibition of 12-O-tetradecanoylphorbol-13-acetate-induced epidermal ornithine decarboxylase activity by phospholipase A_2 inhibitors and lipoxygenase inhibitors. Cancer Res (42): 2841-2845, 1982.
56. Nakadate T, Yamamoto S, Ishii M, Kato R: Inhibition of 12-O-tetradecanoylphorbol-13-acetate-induced epidermal ornithine decarboxylase activity by lipoxygenase inhibitors: possible role of product(s) of lipoxygenase pathway. Carcinogenesis (3): 1411-1414, 1982.

Chapter 4

EFFECTS OF PHORBOL ESTERS ON PHOSPHOLIPID METABOLISM IN VITRO
Philip W. Wertz
University of Iowa
Iowa City, IA 52242

Page

I. Introduction. 74

II. Choline incorporation 75

III. Choline release 86

IV. Arachidonic acid release. 88

V. Phosphatidylethanolamine methylation. 90

VI. Perspectives and unanswered questions 92

References. 94

S.M. Fischer and T.J. Slaga (eds.), ARACHIDONIC ACID METABOLISM AND TUMOR PROMOTION. Copyright © 1985. Martinus Nijhoff Publishing, Boston. All rights reserved.

4. EFFECTS OF PHORBOL ESTERS ON PHOSPHOLIPID METABOLISM IN VITRO
PHILIP W. WERTZ

1. INTRODUCTION

It has long been suspected that due to their lipophilic nature, the tumor promoting phorbol esters exert their primary effect within the cellular membrane system, and the earliest evidence indicating that membranes are in fact targets for phorbol ester action was presented by Sivak and associates (1-3). Investigations prompted by these early reports have revealed that there is a specific phorbol ester binding component which is at least partially localized within cell membranes (4,5) and one of the earliest and most striking consequences of phorbol ester action is the stimulation of membrane phospholipid metabolism. This has proven to be the case both with the in vivo mouse skin model (6-9) and with a variety of cell types in vitro.

Although the in vivo model system is ultimately irreplaceable for studies of chemical carcinogenesis, cells in culture do offer several advantages for mechanistic studies. In general, cell culture model systems are less expensive, require less space and are more amenable to experimental manipulation than their in vivo counterparts. In culture, cells can be synchronously staged at specific points within the cell cycle (10), concentrations of test compounds can be more accurately controlled and extreme conditions such as anaerobiosis can be imposed. Because of this greater flexibility, most studies of the effects of phorbol esters on phospholipid metabolism have been conducted with cells in culture.

The effects of phorbol esters on phospholipid metabolism which have thus far been studied fall into four general areas: choline incorporation, choline release, arachidonic acid release, and phosphatidylethanolamine methylation. Within this chapter, each of these areas will be reviewed; however, phorbol ester stimulated choline incorporation has been the most extensively studied of these several areas and will therefore be discussed in greatest detail. Finally, an attempt will be made to explain the probable and possible relationships among these different aspects of phospholipid metabolism, and to indicate some of the outstanding problems that remain to be solved.

2. CHOLINE INCORPORATION

Süss, Kinzel and Kreibich were the first investigators to demonstrate that phorbol esters stimulate the incorporation of radiolabelled choline into phosphatidylcholine of HeLa cells in culture (11-13), and subsequently Wertz and Mueller further characterized this response with cultured bovine lymph node lymphocytes (14-19). In both systems, the response is elicited by nanomolar concentrations of 12-0-tetradecanoylphorbol-13-acetate (TPA) and is insensitive to the inhibition of RNA or protein synthesis by actinomycin D or cycloheximide respectively (13,14). The relative potencies of a series of phorbol derivatives as stimulators of choline incorporation paralleled their activities as tumor promotors on mouse skin (13,14,20). Phorbol and phorbol triacetate are essentially inactive; phorbol dibutyrate is weakly active, while phorbol didecanoate is nearly as effective as TPA.

The German workers also examined the TPA-responsiveness of a variety of cell lines in addition to HeLa cells (13). These lines included Chang liver cells, CV-1 cells, BALB/c 3T3 cells and BALB/c 3T3 SV transformed cells, all of which respond. One strain of mouse L-cells was found which responds only poorly to the phorbol ester; however, this was shown to be related to the ability of these cells to rapidly metabolize TPA into inactive materials (12).

The bovine lymph node lymphocyte system has proven to be particularly useful. This cell type was chosen because large numbers of cells with relatively low basal metabolic activity are readily obtained. These cells are normally nonreplicative and are locked into a G_O state within the cell cycle; however, they can be activated and induced to undergo hypertrophy and replication by a number of agents including mitogenic lectins (22). In this regard, the phorbol esters are not mitogenic to bovine lymphocytes under the usual conditions of culture; rather, they act as comitogens when used in conjunction with suboptimal doses of T-cell mitogens such as phytohemagglutinin or concanavalin A (23,24). In addition to stimulating phospholipid metabolism and serving as comitogens in the lymphocyte system, the phorbol esters also have been shown to enhance the capping of fluoresceine-conjugated concanavalin A (19,25), to accelerate the transport of glucose (26) and amino acids (27) and to enhance the induction of ornithine decarboxylase in phytohemagglutinin-treated cultures (16,28).

In preliminary experiments[1], the effects of TPA on the uptake and incorporation of $^{32}P_i$ by lymphocytes were examined. It was found that TPA treatment causes an essentially immediate increase in $^{32}P_i$ uptake and enhances the labelling of all phospholipid groups. The increased phospholipid labelling, however, displays biphasic kinetics and cannot be accounted for simply by increased $^{32}P_i$ uptake. Since the largest effect was seen with phosphatidylcholine, the incorporation of ^3H-choline into lipids was next examined (14). In this case, the phorbol ester does not affect choline transport as judged by monitoring cell-associated acid soluble radioactivity. After a lag period of about 20 minutes, TPA stimulates the incorporation of ^3H-choline into phospholipids, and thin-layer chromatographic analysis indicated that phosphatidylcholine, lysophosphatidylcholine and sphingomyeline are all similarly affected. Although phosphatidylcholine is the major one of these three lymphocyte phospholipids and includes the largest portion of the incorporated ^3H-choline, most subsequent measurements, sometimes referred to as phosphatidylcholine labelling, were actually made with unfractionated phospholipids.

Of considerable interest has been the antagonism of

---

1. Wertz, PW, Mueller, GC, unpublished observations.

TPA-enhanced choline incorporation by a variety of agents which also oppose tumor promotion in vivo. This phenomenonological similarity has raised the possibility that the lymphocyte model system is useful for studies on the mechanisms of action and the interactions of both the tumor promoting phorbol esters and a number of antipromoters. Included among the antipromoting agents which inhibit the TPA-stimulation of phospholipid metabolism in lymphocytes are certain protease inhibitors, retinoids and inhibitors of arachidonic acid metabolism. One of the protease inhibitors which has been found to prevent the TPA-mediated acceleration of choline incorporation is N-benzoyl-tyrosine ethyl ester, a chymotrypsin substrate (19). This protease inhibitor not only prevents the phorbol ester-activation of choline metabolism, but prevents or antagonizes all phorbol ester effects in the lymphocyte system including the almost immediate enhancement of concanavalin A capping (19,25). This observation has been taken as an indication of involvement of a protease or esterase at or proximate to the primary phorbol ester-mediated event.

Retinoids which have been studied as antagonists of the TPA-acceleration of choline incorporation in the lymphocyte system include all trans-retinoic acid, 5,6-epoxyretinoic acid, β-ionone and 5,6-epoxy-β-ionone (all shown in Fig. 1). The dose response curve for retinoic acid inhibition of TPA-stimulated choline incorporation is broad, extending from 1 nM through 10 µM (14). Although other interpretations are possible, this broad dose-response curve raised the possibility that retinoic acid may require metabolic activation to produce the active inhibitory species. This observation, combined with the serendipitous finding that certain insect juvenile hormones which are structurally related to epoxyretinoic acid (see Fig. 1) also have retinoid-like activity in the lymphocyte system (28), led to the preparation and testing of several oxygenated retinoids.

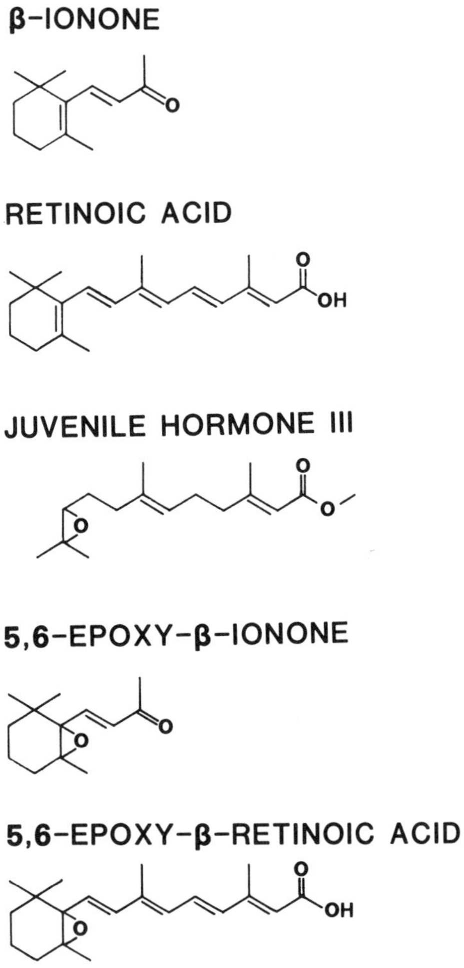

β-IONONE

RETINOIC ACID

JUVENILE HORMONE III

5,6-EPOXY-β-IONONE

5,6-EPOXY-β-RETINOIC ACID

FIGURE 1. Retinoic acid and structurally related substances.

First, 5,6-epoxy-β-ionone was studied (15). Although β-ionone itself was found to have no inhibitory activity against TPA-stimulation of choline incorporation, the epoxide is active in the 10-300 μM range. While the potency of this antagonist is not great, it encouraged the synthesis and testing of 5,6-epoxyretinoic acid (16). This retinoid proved to be of comparable activity to retinoic acid as an inhibitor of the phorbol ester-acceleration of choline incorporation. Most interestingly, it is capable of completely blocking the induction of ornithine decarboxylase activity by TPA under conditions where the parent retinoic acid is inactive. Retinoic acid opposes this induction only if it is added to lymphocyte cultures along with phytohemagglutinin one hour prior to TPA. The 5,6-epoxide is effective when added to cultures concurrently with TPA.

At about the same time that synthetic 5,6-epoxyretinoic acid was found to be biologically active in bovine lymphocytes, DeLuca and associates at the University of Wisconsin discovered that the 5,6-epoxide is a major polar metabolite of retinoic acid in the intestinal mucosa of the rat (29,30). This raised considerable interest in epoxyretinoic acid and led to its prompt examination in the two-stage mouse skin carcinogenesis system (31). Within this model, 5,6-epoxyretinoic acid was found to have antipromotor activity comparable to but not greater than all trans-retinoic acid when the TPA-inductions of ornithine decarboxylase, papillomas and carcinomas were monitored. Furthermore, although less potent, 5,6-dihydroretinoic acid also displays anti-TPA activity. Since this material would not readily be converted to the 5,6-epoxy derivative, interest in the anticarcinogenic activity of the oxyretinoid dwindled. This is perhaps unfortunate since there are relatively few other agents which are as potent as retinoic acid in opposing tumor promotion. Possible differences in the pharmacokinetics, mechanisms of action and relative toxicities of 5,6-epoxyretinoic acid and all trans-retinoic acid remain largely unknown. Such differences could be of basic scientific interest as well as practical importance.

Inhibitors of arachidonic acid metabolism were also tested as modifiers of the phorbol ester-activation of phosphatidylcholine labelling (17). The standard fatty acid cyclooxygenase inhibitor, indomethacin, was found to inhibit choline incorporation in TPA-treated lymphocytes; however, this inhibition is not selective for the phorbol ester effect, requires concentrations in the millimolar range, and cannot be reversed by exogenously supplied arachidonic acid, prostaglandins or prostacyclins (17). Also, the selective thromboxane synthetase inhibitor, 9,11-azoprosta-5,13-dienoic acid (32), has no effect on choline incorporation. These results indicate that fatty acid cyclooxygenase products are not involved in the activation of choline phospholipid metabolism by TPA.

In contrast to this behavior, 5,8,11,14-eicosatetraynoic acid (ETYA) selectively inhibits the effect of TPA on choline incorporation at concentrations in the 1-30 µM range, and this inhibitory action can be precluded by concurrently added arachidonic acid (17). ETYA is the acetylenic analog of arachidonic acid (33) and inhibits both the fatty acid cyclooxygenase and lipoxygenase pathways (34,35). Its mode of inhibition includes both reversible competitive and time dependent irreversible components. The latter inhibitory mode may arise when a hydrogen is abstracted from a methylene group adjacent to a triple bond; the resulting allylic free radical is highly reactive and presumably alkylates amino acid residues at or near the active site of the enzyme (36). As noted above, arachidonic acid added to lymphocyte cultures at the same time as ETYA prevents the inhibition of accelerated choline incorporation; however, if the addition of arachidonic acid to culture is delayed, the inhibitory effect of ETYA becomes progressively less reversible (37). This presumably reflects the time-dependent irreversible mode of ETYA action. As in the case of indomethacin-inhibited choline incorporation, ETYA-inhibition cannot be prevented or reversed by exogenously supplied prostaglandins (types A_1, A_2, B_1, B_2, D_2, E_1, E_2, $F_{1\alpha}$ and $F_{2\alpha}$) or prostacyclins (PGI_1 and 6,9-thia-PGI_2). In fact, the type E prostaglandins and prostacyclins significantly enhance ETYA-inhibition. By the process of elimination, a product of the lipoxygenase pathway or some similar oxidative pathway would seem to be involved in the TPA-activation of choline incorporation. In support of this suggestion, the phorbol ester effect has been shown to depend upon the presence of oxygen (17,19,37).

The concept of phorbol ester action which follows from these studies includes an initial phorbol ester action which leads secondarily to the conversion of a preexisting but relatively inactive choline incorporation system to a more fully active state. This conversion would seem to involve an oxygen dependent ETYA sensitive mechanism. Several lines of investigation have followed from this interpretive view and these are summarized below.

One obvious avenue of research was to look for the production of arachidonic acid metabolites in TPA-treated lymphocytes, and this has now been done (38). Two such metabolites have been found. The system which produces these metabolites is fully active within 10 minutes of TPA-treatment; also, it requires oxygen and can be inhibited by ETYA and retinoic acid. At low concentrations, indomethacin enhances production of these metabolites. Nordihydroguaiaretic acid, another lipoxygenase inhibitor, also antagonizes the TPA-mediated production of both metabolites. The identities of these arachidonic acid metabolites are yet to be determined; however, they do not appear to correspond to any of the known prostanoids. Their mobility on silicic acid thin-layer chromatography is intermediate between that of the least polar phospholipids and 6-keto-$PGF_{1\alpha}$. Other prostaglandins, HETE's and related metabolites were all less polar and were produced at comparable rates in both control and TPA-treated lymphocytes. The TPA-sensitive metabolites are not detectable in control cells.

A second possibility which has been considered is that a reactive lipoxidase product covalently modifies proteins and thereby alters catalytic activities including choline incorporation. Such a process would not be evident from analysis of the lipid extracts; rather, the macromolecules or protein fraction should be examined. Toward this end, lymphocytes were treated with different doses of TPA in the presence of ^{3}H-arachidonic acid, and after a two hour incubation the cells were acid precipitated and extensively lipid extracted. The residues were still radiolabelled, and the extent of labelling was dependent upon the dose of TPA to which the cells had been exposed (37). Much of this label fractionated along with the proteins during phenol-water partitioning. This protein associated radioactivity was nondialyzable but was rendered dialyzable by pronase digestion. Although preliminary experiments indicated that the label may be covalently attached to proteins through thioether linkages, subsequent electrophoretic analysis failed to demonstrate any specifically labelled coomassie blue staining bands. Although it does not seem probable, the possibility of tightly bound low molecular weight arachidonic acid derived materials cannot be ruled out. More likely, the radioactivity is covalently attached to proteins, but is distributed over many species so that the specific activity of any one is low.

Still another area of interest has been the enzymology underlying the incorporation of choline into phospholipids and the modifications brought about by phorbol ester action upon this metabolic machinery. The principal pathway by which choline is incorporated into phospholipids was elucidated by E. P. Kennedy (39) and is summarized in Figure 2. Within this pathway, choline is first phosphorylated at the expense of one ATP to produce phosphorylcholine. The phosphorylcholine is then condensed with CTP to produce CDP-choline and pyrophosphate. Finally, a choline phosphate is transferred from CDP-choline to diacylglycerol to produce phosphatidylcholine or to a ceramide to produce sphingomyelin. The enzymes of this pathway have been studied most extensively by Schneider (40-42) and by Vance (43-50) and their associates, and it is now known that the rate determining step is the formation of CDP-choline (45), which is catalyzed by CTP:phosphorylcholine cytidyltransferase.

In bovine lymphocytes, TPA-treatment has been shown to enhance the activity of cytidyltransferase which can be measured in cell homogenates (19,37,51). In control cells, cytidyltransferase activity is largely confined to the 100,000 x g particulate fraction, but there is also a low level of soluble activity (37,51). After TPA-treatment the increase in activity is found entirely within the particulate fraction; the low soluble activity is further reduced. The phorbol ester effect cannot be produced by adding TPA directly to lymphocyte sonicates; however, cytidyltransferase can be activated by addition of arachidonic acid to cell sonicates. This activation by arachidonic acid does not occur with the soluble lymphocyte fraction in the absence of particulate material.

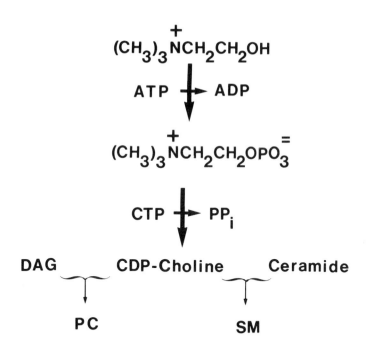

FIGURE 2. The metabolic pathway for incorporation of choline into phospholipids (39). Diacylglycerol (DAG), phosphatidylcholine (PC), sphingomyelin (SM).

More recently, Vance and associates have studied the effect of TPA on phosphatidylcholine synthesis in HeLa cells (48,49). First, analysis of the acid soluble choline metabolites from intact HeLa cells demonstrated that TPA treatment does increase the rate of conversion of phosphorylcholine to CDP-choline (48). After a one hour pulse labelling with ^{3}H-choline, most of the water soluble label is associated with phosphorylcholine. Treatment with TPA accelerates the conversion of labelled phosphorylcholine into phosphatidylcholine, while CDP-choline associated label remains low and relatively constant throughout the experiment. These results confirm that the formation of CDP-choline limits the incorporation rate of choline into phospholipids while the phosphorylation of choline and the transfer of phosphorylcholine from CDP-choline to diacyl-glycerol are rapid steps. The data strongly argue that TPA accelerates CDP-choline formation.

When cytidyltransferase activity was measured in homogenates of control and TPA-treated cells, the latter were found to have four-fold greater activity[2]. Upon subcellular fractionation of the homogenates, the TPA increased cytidyltransferase activity was found to be confined to the microsomal fraction (49); activity in the cytosolic fraction decreases as a consequence of phorbol ester action. Therefore, the TPA-activation of cytidyltransferase is accompanied by translocation of the enzyme to a membranous location. Additional evidence for the particulate nature of the activated cytidyltransferase was obtained by exposing control and TPA-activated cells to digitonin[2]. This drug binds to membrane cholesterol and causes the cells to become permeable to soluble cytidyltransferase. It was observed that cytidyltransferase activity is released at a much reduced rate from cells which have undergone phorbol ester treatment[2]. Thus, the argument that phorbol ester action converts relatively inactive soluble cytidyltransferase units into more active membrane associated or aggregated complexes is further supported.

---

2. Personal communication from DE Vance and SL Pelech.

Pelech and Vance have also made observations implicating phospholipases of the A and C types in the mechanism by which phorbol esters activate choline incorporation in HeLa cells (49). First, treatment with phospholipase C mimics the action of TPA on cytidyltransferase translocation, activation, and phosphatidylcholine synthesis. Also, free fatty acids, which are products of phospholipase A_2 action, have been found to bind to soluble cytidyltransferase units and to facilitate the transfer of these units onto membranes in vitro (49,50). Fatty acids do not activate the cytidyltransferase; activation occurs after translocation to the membranes^2. The calcium ionophore A23187, which may serve to activate phospholipase A_2, stimulates the incorporation of choline into phospholipids by increasing the rate of the cytidyltransferase reaction^2, and a number of phospholipase inhibitors antagonized the phorbol ester stimulation of choline incorporation in HeLa cells (49). These inhibitors include trifluoperazine, chlorpromazine, dibucaine, tetracaine and mepacrine.

The picture of phorbol ester mediated acceleration of choline incorporation which emerges from these observations includes the following: phorbol ester binding results in the activation of phospholipases which may produce among their products arachidonic acid and other free fatty acids, lysophospholipids and diacylglycerol. This initial esterase-mediated phase of phorbol ester action may be associated with a Ca^{++} flux. The free fatty acids liberated from membrane phospholipids bind to the soluble cytidyltransferase units and in doing so may create a lipophilic surface which serves in the transfer and binding of the enzyme to the membrane system. Once bound to membranes, the enzyme is activatable, and potential activators include acidic phospholipids (43,44), lysophospholipids (42,44), oxidized phospholipids (42), and arachidonic acid derived substances (17,19,37). Results obtained with the bovine lymphocyte system would point most strongly toward this latter group of substances as the most likely activators; however, the cytidyltransferase is a complicated regulatory enzyme and may be subject to the influence of multiple regulatory substances. The relative importance of these various factors may be a function of the state of differentiation of the cell under study or the conditions of culture. One recently reported complication has indicated that the cytidyltransferase in rat liver cytosol is a phosphorylated enzyme of low activity (51). Removal of phosphates or binding of free fatty acids promote binding of the enzyme to microsomes where activation takes place.

3. CHOLINE RELEASE

Once again, Kinzel, Kreibich, Hecker and Süss, who studied the stimulation of choline incorporation by phorbol esters, also noted that TPA stimulates the release of radioactivity into the medium from HeLa cells which had been prelabelled with ^{3}H-choline (13), but the nature of the released radioactive material was not determined. This reaction has subsequently been investigated in greater detail, principally by Weinstein and associates (52-54), by Guy and Murray (55,56), and by Grove and Schimmel (57,58). Mufson, Odin and Weinstein, working with C3H10T½ mouse embryo fibroblasts which were prelabelled with ^{3}H-choline, found that TPA stimulates release of radioactivity into the medium twofold within 5 minutes (52). The reaction continues for at least 2 hours, by which time the released radioactivity from TPA treated cells is as much as fivefold that found in control cultures. TPA, phorbol didecanoate and mezerein are all equally effective in stimulating the release reaction when tested at 10^{-7}M; phorbol dibutyrate is weakly active and 4-α-phorbol didecanoate is inactive. The choline-release reaction is temperature dependent but does not depend upon RNA or protein synthesis as judged by experiments with cordycepin and cycloheximide (52). Also, choline release is not activated by Ca^{++} ionophore A23187 and is insensitive to ETYA (52).

The released radioactivity was analyzed by paper chromatography and found to be associated with choline and phosphorylcholine (52). Paddon and Vance found that phosphorylcholine is released into the medium during TPA-treatment of HeLa cells

(48), while Guy and Murray reported release of both choline and phosphorylcholine (55). The phosphorylcholine is presumably released by phospholipase C activity while choline would be released by a phospholipase D. However, type-D phospholipases have not been identified in animal tissues. An alternative mechanism for the release of choline would be the action of phospholipase C followed by hydrolysis of the resulting phosphorylcholine.

Guy and Murray presented evidence that the extruded label is not simply a result of cell leakiness (55). It was observed that TPA does not induce the release of radioactivity from HeLa cells prelabelled with ^{14}C-thymidine or ^{14}C-hypoxanthine. Furthermore, the TPA-enhanced release of radiolabelled choline and phosphorylcholine reflects more closely the radiolabelling of the phospholipids rather than the acid soluble pool.

Working with HeLa cells, Guy and Murray also demonstrated that TPA-activation of choline release clearly precedes the stimulation of choline incorporation (55). It was suggested that TPA-activation of a phospholipase C may be a primary event which causes choline release and which ultimately leads to the stimulation of phosphatidylcholine resynthesis. In support of this view, it was shown that treatment with exogenous phospholipase C results in the stimulation of choline incorporation.

Interestingly, the phospholipase C from Clostridium perfringens is active while the same level of activity from Bacillus cereus is not (55). Also, treatment with a phospholipase D fails to release radioactivity. This latter observation would indicate that either the enzyme is not active under the conditions of the assay or, more interestingly, the newly labelled phosphatidylcholine is concentrated within the inner leaflet of the plasma membrane or other internal cell membranes. Additional work would be required to clarify this point.

In contrast to the results of Mufson et al (52), Guy and Murray found that the Ca^{++} ionophore A23187 does enhance choline release both from control and TPA-treated HeLa cells (55). Triflouperazine antagonizes the stimulation of choline-release by either TPA or A23187.

Finally, Grove and Shimmel demonstrated that TPA-treatment of differentiated chick embryo myoblasts results in a two-fold increase in the level of 1,2-diacylglycerol within 15-30 min (57,58). This diacylglycerol pool probably arises from phospholipase C action on phosphatidylcholine as judged by similarities in fatty acid composition and rates of glycerol turnover. The release reaction in this case was found to be independent of the level of extracellular Ca^{++}, cycloheximide, indomethacin and La^{+++}. The La^{+++} ion is similar to Ca^{++} and interferes with Ca^{++} uptake.

So it appears that an early consequence of phorbol ester interaction with cells is the activation of a phosphatidylcholine specific phospholipase C. The action of this process leads to the liberation of phosphorylcholine and choline into the extracellular medium and the formation of diacylglycerol within the cell. Weinstein has suggested a scheme in which diacylglycerol, produced by phospholipase C, is subsequently acted upon by a diacylglycerol lipase to liberate arachidonic acid, which in turn serves as precursor for production of the various prostanoids (53,54). While this possibility remains to be tested, the liberation of fatty acids by phospholipase A_2 in TPA treated cells has been studied. This area of study is summarized below.

4. ARACHIDONIC ACID RELEASE

The action of phospholipase A_2 is removal of the fatty acid from the sn-2 position of phospholipids. This reaction produces free arachidonic acid from which prostaglandins and other prostanoids are produced. As cellular free fatty acid pools are generally very small, phospholipase A_2 is often thought to determine the rate of prostanoid production. Prostaglandins and related metabolites are now known to be involved in phorbol ester-induction of ornithine decarboxylase activity (59,60), cell proliferation (61) and tumor promotion in vivo (60,62); therefore the liberation of arachidonic acid from cellular lipids is of some interest. Levene and Hassid were the first to demonstrate that phorbol esters stimulate the release of arachidonic acid from Madin-Darbey canine kidney (MDCK) cells (63). Subsequently, Daniel, King and Waite demonstrated that phosphatidylethanolamine is the principal source of arachidonic acid in MDCK cells, and when cells are labelled with both ^{3}H-arachidonic acid and ^{14}C-palmitic acid, TPA-treatment causes preferential release of arachidonic acid (64). This presumably reflects the preferential concentration of arachidonic acid in the <u>sn</u>-2 position and the activation of a phospholipase A_2. Consistent with this suggestion, Tou has demonstrated that TPA-treated granulocytes show an immediate stimulation of fatty acid turnover, and the effect is greatest at the <u>sn</u>-2 position (65). Thus, it appears that phorbol esters increase a phospholipase A_2 activity, and the lysophosphatide product is efficiently reacylated. In the case of MDCK cells, arachidonic acid is liberated, and saturated and monoenoic fatty acids replace it in the phospholipid structure (64).

TPA-stimulated arachidonic acid release from MDCK cells was found to be inhibited by cycloheximide (66), high concentrations of indomethacin (66) and α-tocopherol (67). The inhibition by α-tocopherol appears to involve interference with phorbol ester binding, but the mechanism underlying this effect is uncertain.

Mufson, DeFeo and Weinstein have studied arachidonic acid release from chick embryo fibroblasts (68). Nanomolar concentrations of TPA cause release of arachidonic acid and formation of prostaglandins within one hour. The phorbol ester effect is inhibited by cycloheximide and puromycin but is not sensitive to actinomycin D. The possibility was suggested that the apparent requirement for protein synthesis may relate to secretion rather than to esterase action.

It was also found that all trans-retinoic acid is an effective inhibitor of the TPA-enhancement of arachidonic acid release in the chick embryo fibroblast system during a 3 hour incubation (68). Retinoids have no effect on arachidonic acid release from control cells. This contrasts with the findings of Ohuchi and Levine, who reported that retinoids enhance the release of arachidonic acid from MDCK cells (69). These workers found that all trans-retinoic acid, 13-cis-retinoic acid, retinal, retinol, retinyl acetate, retinyl palmitate and the trimethylmethoxyphenyl analog (TMMP) all stimulate fatty acid release and prostaglandin production during a 22 hour incubation. All trans-retinoic acid is the most effective. The retinoids do not antagonize phorbol ester enhancement of arachidonic acid release in this system. In fact, TMMP acts synergistically when used in combination with TPA.

The acute effects of phorbol esters on arachidonic acid release from rabbit peritoneal leukocytes have been examined by Axelrod and associates (70). In these cells, TPA enhances the release of arachidonic acid from phospholipids within 1-2 minutes. The release reaction continues for at least 20 min., and was found to be associated with the degradation of a phosphatidylcholine pool which is synthesized by methylation of phosphatidylethanolamine (71). This methylation pathway is discussed in the following section.

5. PHOSPHATIDYLETHANOLAMINE METHYLATION

It has recently been documented that the conversion of phosphatidylethanolamine to phosphatidylcholine via successive methylation, although a quantitatively minor pathway (71,72), is of significance in the transduction of signals across membranes (73). This biochemical process involves the successive transfer of three methyl groups from S-adenosylmethionine to the phosphatidylethanolamine nitrogen as shown in Figure 3. The first methyl transfer to produce phosphatidyl-N-monomethyl-ethanolamine is catalyzed by a methyl transferase which is located on the cytoplasmic side of the plasma membrane (74). The monomethyl derivative is catalytically transferred from inner to outer leaflet of the plasma membrane where conversion to phosphatidylcholine is completed through the action of a second methyl transferase (74). It has been proposed that a variety of different types of cell surface receptors may be associated with the phospholipid methylation system to form "signal clusters" (73). Binding of ligand to its receptor would activate the methylation pathway and alter the membrane microenvironment. Stimulation of phospholipid methylation in a number of systems has been associated with increases in Ca^{++} uptake and arachidonic acid release. Ca^{++} is a phospholipase A_2 activator, and as noted above, there is some evidence that the phosphatidylcholine produced by the methylation route is arachidonic acid rich and serves as a preferential substrate for a phospholipase A_2 (70,73).

PHOSPHOLIPID METHYLATION PATHWAY

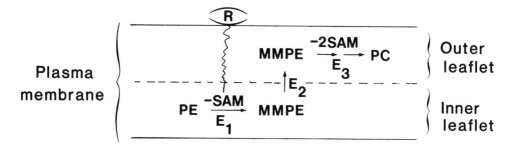

FIGURE 3. Conversion of phosphatidylethanolamine (PE) to phosphatidylcholine (PC) at the expense of 3S-adenosylmethionines (SAM). R represents a receptor coupled to E_1, the inner leaflet methyltransferase which converts PE to monomethylphosphatidylethanolamine (MMPE). MMPE is catalytically transferred to the outer leaflet where conversion to PC is completed through the action of E_3.

The effects of phorbol esters on the methylation pathway have thus far been examined in only one cell type--the HL-60 promyelocytic leukemia cell line. TPA-treatment of these leukemia cells induces them to undergo differentiation and to acquire morphological, cytochemical and biochemical characteristics of macrophages (75,76). Huberman and associates have detected a number of alterations in neutral lipid and phospholipid metabolism which occur during this phorbol ester induced process, and enhanced phospholipid methylation is among the detected changes (77,78). TPA enhances phospholipid methylation detectably within 6 hours, and the effect is maximum within 24 to 48 hours (78). The methylation response is dependent upon the dose of TPA and is not produced by 4-0-methyl-TPA or phorbol-12,13-diacetate, both of which also fail to induce differentiation in this cell line. TPA-enhanced methylation reflects an increase in methyltransferase activity which was measured in cell lysates. One HL-60 variant was isolated which is resistant to phorbol esters as inducers of differentiation; TPA does not stimulate phospholipid methylation in these cells.

In contrast to these results, Cassileth, Suholet and Cooper found that tumor-promoting phorbol esters inhibit phospholipid methylation in HL-60 leukemia cells (78). This inhibition is dependent upon the dose of TPA, and the structure-activity relationships for a series of phorbol derivatives reflects the effectiveness of these compounds as inducers of differentiation in HL-60 cells or as tumor promoters on mouse skin. Decreased methylation correlates with increased incorporation of choline via the CDP-choline pathway.

The differences which these two groups have reported for phosphatidylethanolamine methylation in TPA-treated HL-60 cells may reflect differences in the variants of the cell line which were studied. What is most impressive in both sets of experiments is the fact that phorbol esters do modulate the phospholipid methylation reaction. These observations should be sufficient to stimulate other researchers in the area of carcinogenesis to consider and to explore the possible relationships of phospholipid methylation to tumor promotion.

6. PERSPECTIVES AND UNANSWERED QUESTIONS

Although phorbol esters appear to exert a variety of effects on phospholipid metabolism, all of these effects probably stem from interaction of the phorbol ester with a common receptor. Furthermore, it is possible that more than one of these effects is a direct consequence of the phorbol ester binding to its receptor; a single type of receptor molecule may be coupled to different signal transducers at different sites within the membrane of one cell or may be coupled to different arrays of transducers in different cell types. Among the best candidates for transducers coupled to phorbol ester receptors are phospholipase C and the phospholipid methyltransferases. Although TPA-enhanced phospholipase A_2 activity appears to reflect an inductive process in some cells, in other cells this activity too may be directly coupled to the receptor.

One common consequence of the actions of these potential transducers -- phospholipase C, phospholipase A_2 and the methyltransferases -- is a localized modification of the physical properties of the membrane. Such membrane modification may have profound effects upon membrane associated enzymes and may result in Ca^{++} fluxes and a variety of secondary metabolic events.

A second common aspect of phospholipase A_2, phospholipase C or methyltransferase action is the mobilization of arachidonic acid and other free fatty acids. In the case of phospholipase A_2, free fatty acids including arachidonic acid are direct products. In the case of phospholipase C, the initially produced diacylglycerol may serve as a substrate for a diglyceride lipase. And the methyltransferase system appears to convert a localized arachidonic acid rich pool of phosphatidylethanolamine into phosphatidylcholine, which is subsequently hydrolyzed by a phospholipase A_2. The free fatty acids released by these reactions seem to facilitate the translocation of a soluble CTP:phosphorylcholine cytidyltransferase from the soluble compartment to the membrane system. Once membrane bound, this cytidyltransferase becomes activated, possibly by an oxidative metabolite of arachidonic acid. Other arachidonic acid metabolites serve as mediators of a variety of processes in the biochemical cascade which ultimately leads to modulation of gene expressions.

There are many important questions remaining concerning the role of lipid metabolism in tumor promoter action. For one thing, the area of arachidonic acid metabolism, although much studied in recent years, is still incompletely understood. For instance, what are the structures of the unusual metabolites reported by Wrighton, Pai and Mueller? Will these materials prove to be activators of cytidyltransferase? Do they represent a new class of lipoxygenase products, or are they possibly produced by some entirely new oxygen dependent, ETYA-sensitive process?

What about the other lipid-derived materials which are now known to be produced by phorbol ester action? What is the functional significance of extracellular phosphorylcholine or of the accumulation of lysophospholipids or diacylglycerols? By what mechanisms do retinoids interfere with phospholipid deacylation or choline incorporation? Are all retinoic acid effects mediated by retinoic acid binding proteins, or are some retinoic acid metabolites capable of modifying cellular metabolism more directly?

I am certain that many of these questions are currently under investigation. Some will probably be answered by the time this volume appears in print, but it is hoped that this review will stimulate new investigations in the area of lipid metabolism and tumor promotion.

REFERENCES

1. Sivak A, Ray F, Van Duuren BL: Phorbol ester tumor promoting agents and membrane stability. Cancer Res (29):624-630, 1969.
2. Sivak A, Mossman BT, Van Duuren BL: Activation of cell membrane enzymes in the stimulation of cell division. Biochem Biophys Res Commun (46):605-609, 1972.
3. Sivak A: Induction of cell division: role of cell membrane sites. J Cell Physiol (80):167-174, 1972.
4. Dreidger PE, Blumberg PM: Specific binding of phorbol ester tumor promoters to mouse skin. Proc Natl Acad Sci USA (77):567-571, 1980.
5. Shoyab M, Todaro GJ: Specific high affinity cell membrane receptors for biologically active phorbol and ingenol esters. Nature (288):451-455, 1980.
6. Kreibich G, Hecker E, Süss R, Kinzel V: Phorbol ester stimulates choline incorporation. Naturwissenschaften (58):323, 1971.

7. Rohrschneider LR, O'Brien DH, Boutwell RK: The stimulation of phospholipid metabolism in mouse skin following phorbol ester treatment. Biochim Biophys Acta (280):57-70, 1972.
8. Rohrschneider LR, Boutwell RK: The early stimulation of phospholipid metabolism by 12-0-tetradecanoylphorbol-13-acetate and its specificity for tumor promotion. Cancer Res (33):1945-1952, 1973.
9. Balmain A, Hecker E: On the biochemical mechanism of tumorigenesis in mouse skin. VI. Early effects of growth stimulating phorbol esters on phosphate transport and phospholipid synthesis in mouse epidermis. Biochim Biophys Acta (362):457-469, 1974.
10. Mueller GC: Biochemical perspectives of the G_1 and S intervals in the replication cycle of animal cells: A study in the control of cell growth. In: Baserga R (ed) The cell cycle and cancer. Marcel Dekker Inc., New York, 1971, pp 269-307.
11. Süss R, Kinzel V, Kreibich G: Cocarcinogenic croton oil factor A_1 stimulates lipid synthesis in cell cultures. Experientia (27):46-47, 1971.
12. Süss R, Kreibich G, Kinzel V: Phorbol esters as a tool in cell research? Eur J Cancer (8):299-304, 1972.
13. Kinzel V, Kreibich G, Hecker E, Süss R: Stimulation of choline incorporation in cell cultures by phorbol derivatives and its correlation with their irritant and tumor promoting activity. Cancer Res (39):2743-2750, 1979.
14. Wertz PW, Mueller GC: Rapid stimulation of phospholipid metabolism in bovine lymphocytes by tumor-promoting phorbol esters. Cancer Res (38):2900-2904, 1978.
15. Wertz PW, Kensler TW, Mueller GC: Inhibition of phorbol ester action in lymphocytes by 5,6-epoxy-β-ionone. Biochem Biophys Res Commun (83):138-143, 1978.
16. Wertz PW, Kensler TW, Mueller GC, Verma AK, Boutwell RK: 5,6-Epoxyretinoic acid opposes the effects of 12-0-tetradecanoylphorbol-13-acetate in bovine lymphocytes. Nature (277):227-229, 1979.
17. Wertz PW, Mueller GC: Inhibition of 12-0-tetradecanoyl-phorbol-13-acete-accelerated phospholipid metabolism by 5,8,11,14-eicosatetraynoic acid. Cancer Res (40): 776-781, 1980.
18. Wertz PW, Mueller GC: Activation of CTP:phosphorylcholine cytidyltransferase by 12-0-tetradecanoylphorbol-13-acetate. Proc Amer Assoc Cancer Res (21):511, 1980.
19. Mueller GC, Wertz PW, Kwong CH, Anderson KA, Wrighton SA: Dissection of the early molecular events in the activation of lymphocytes by 12-0-tetradecanoylphorbol-13-acetate. Jerusalem Symposia on Quantum Chemistry and Biochemistry (13):319-333, 1980.
20. Boutwell RK: The function and mechanism of promoters of carcinogenesis. CRC Crit Rev Toxicol (2):419-443, 1974.
21. Kreibich G, Süss R, Kinzel V: On the biochemical mechanism of tumorigenesis in mouse skin. V. Studies of the metabolism of tumor promoting and non-promoting phorbol derivatives in vivo and in vitro. Z. Krebsforsch (81): 135-149, 1974.

22. Nowell PC: Phytohemagglutinin: An initiator of mitosis in cultures of normal human leukocytes. Cancer Res (20):462-466, 1960.
23. Mastro AM, Mueller GC: Synergistic action of phorbol esters in mitogen-activated bovine lymphocytes. Exptl Cell Res (88):40-46, 1974.
24. Kensler TW, Mueller GC: Retinoic acid inhibition of the cornitogenic action of meerein and phorbol esters in bovine lymphocytes. Cancer Res (38):771-775, 1978.
25. Kwong CH, Mueller GC: Antagonism of concanavalin A capping in phorbol ester-activated lymphocytes by calmodulin inhibitors and certain amino acid esters. Cancer Res (42):2115-2120, 1982.
26. Wrighton SA, Mueller GC: Rapid acceleration of deoxyglucose transport by phorbol esters in bovine lymphocytes. Carcinogenesis (3):1415-1418, 1982.
27. Kensler TW, Wertz PW, Mueller GC: Inhibition of phorbol ester-accelerated amino acid transport in bovine lymphocytes. Biochim Biophys Acta (585):43-52, 1979.
28. Kensler TW, Verma AK, Boutwell RK, Mueller GC: Effects of retinoic acid and juvenile hormone on the induction of ornithinedecarboxylase activity by 12-0-tetradecanoyl phorbol-13-acetate. Cancer Res (38):2896-2899, 1978.
29. McCormick AM, Napoli JL, Schnoes HK, DeLuca HF: Isolation and identification of 5,6-epoxyretinoic acid: a biologically active metabolite of retinoic acid. Biochemistry (17):4085-4090, 1978.
30. McCormick AM, Napoli JL, Yoshizawa S, DeLuca HF: 5,6-Epoxyretinoic acid is a physiological metabolite of retinoic acid in the rat. Biochem J (186):475-481, 1980.
31. Verma AK, Slaga TJ, Wertz PW, Mueller GC, Boutwell RK: Inhibition of skin tumor promotion by retinoic acid and its metabolite 5,6-epoxyretinoic acid. Cancer Res (40):2367-2371, 1980.
32. Gorman RR, Bundy GL, Peterson DC, Sun FF, Miller OV, Fitzpatrick FA: Inhibition of human platelet thromboxane synthetase by 9,11-azoprosta-5,13-dienoic acid. Proc Natl Acad Sci USA (74):4007-4011, 1977.
33. Downing DT, Barve JA, Gunstone FD, Jacobsberg FR, Jie MLK: Structural requirements of acetylenic fatty acids for inhibition of soybean lipoxygenase and prostaglandin synthetase. Biochim Biophys Acta (280):343-347, 1972.
34. Ahern DG, Downing DT: Inhibition of prostaglandin biosynthesis by eicosa-5,8,11,14-tetraynoic acid. Biochim Biophys Acta (210):456-461, 1970.
35. Downing DT: Differential inhibition of prostaglandin synthetase and soybean lipoxygenase. Prostaglandins (1):437-441, 1972.
36. Downing DT, Ahern DG, Bachta M: Enzyme inhibition by acetylenic compounds. Biochem Biophys Res Commun (40):218-223, 1970.
37. Mueller GC, Wertz PW: A possible role for protein alkylation in phorbol ester action. In: Hecker E (ed) Carcinogenesis. Raven Press, New York, 1982, Vol. 7, pp 499-511.

38. Wrighton SA, Pai JK, Mueller GC: Demonstration of two unique metabolites of arachidonic acid from phorbol ester-stimulated bovine lymphocytes. Carcinogenesis, in press, 1983.
39. Kennedy EP: The metabolism and function of complex lipids. The Harvey Lectures (57):143-171, 1962.
40. Schneider WC, Behki RM: Phosphorous compounds in animal tissues. VII. Enzymatic formation of deoxycytidine diphosphate choline and lecithin by tissue homogenates. J Biol Chem (238):3565-3571, 1963.
41. Schneider WC: Intracellular distribution of enzymes. XIII. Enzymatic synthesis of deoxycytidine diphosphate choline and lecithin in rat liver. J Biol Chem (238):3572-3578, 1963.
42. Fiscus WG, Schneider WC: The role of phospholipids in stimulating phosphorylcholine cytidyltransferase activity. J Biol Chem (241):3324-3330, 1966.
43. Choy PC, Lim PH, Vance DE: Purification and characterization of CTP:cholinephosphate cytidyltransferase from rat liver cytosol. J Biol Chem (252):7673-7677, 1977.
44. Choy PC, Vance DE: Lipid requirements for activation of CTP:phosphorylcholine cytidyltransferase. J Biol Chem (353):5163-5167, 1978.
45. Vance DE, Choy PC: How is phosphatidylcholine biosynthesis regulated? Trends Biochem Sci (4):145-148, 1979.
46. Vance DE, Trip EM, Paddon HB: Poliovirus increases phosphatidylcholine biosynthesis in HeLa cells by stimulation of the rate-limiting reaction catalyzed by CTP:phosphocholine cytidyltransferase. J Biol Chem (255):1064-1069, 1980.
47. Choy PC, Paddon HB, Vance DE: An increase in cytoplasmic CTP:phosphocholine cytidyltransferase in poliovirus-infected HeLa Cells. J Biol Chem (255):1070-1073, 1980.
48. Paddon HB, Vance DE: Tetradecanoylphorbol acetate stimulates phosphatidylcholine biosynthesis in HeLa cells by an increase in the rate of the reaction catalyzed by CTP:phosphocholine cytidyltransferase. Biochim Biophys Acta (620):636-640, 1980.
49. Pelech SL, Paddon HB, Vance DE: Mechanism of TPA-induced stimulation of phosphatidylcholine biosynthesis in HeLa cells. Fed Proc (42):643, 1983.
50. Vance DE, Pelech SL: Regulation of phosphatidylcholine biosynthesis by reversible phosphorylation and fatty acids. Fed Proc (42):1739, 1983.
51. Wertz PW, Mueller GC: Activation of CTP:phosphorylcholine cytidyltransferase by 12-0-tetradecanoylphorbol-13-acetate. Proc Amer Assoc Cancer Res (20):949, 1979.
52. Mufson RA, Okin E, Weinstein IB: Phorbol esters stimulate the rapid release of choline from prelabelled cells. Carcinogenesis (2):1095-1102, 1981.
53. Weinstein IB: Current concepts and controversies in chemical carcinogenesis. J Supramol Structure and Cell Biochem (17):99-120, 1981.
54. Weinstein IB, Horowitz AD, Mufson RA, Fisher PB, Ivanovic V, Greenebaum E: Results and speculations related to

recent studies on mechanisms of tumor promotion. In: Hecker E (ed) Carcinogenesis, Raven Press, New York, 1982, Vol 7, pp 599-616.
55. Guy GR, Murray AW: Tumor promoter stimulation of phosphatidylcholine turnover in HeLa cells. Cancer Res (42):1980-1985, 1982.
56. Guy GR, Murray AW: Modulation of HeLa cell phospholipid metabolism and ornithine decarboxylase activity by tumor promotors: Regulation of response to phorbol-12,13-dibutyrate by receptor. Biochem Biophys Res Commun (106):1398-1404, 1982.
57. Grove RI, Schimmel SD: Generation of 1,2-diacylglycerol in plasma membranes of phorbol ester-treated myoblasts. Biochem Biophys Res Commun (102):158-164, 1981.
58. Grove RI, Schimmel SD: Effects of 12-0-tetradecanoyl-phorbol-13-acetate on glycerolipid metabolism in cultured myoblasts. Biochim Biophys Acta (711):272-280, 1982.
59. Verma AK, Rice HM, Boutwell RK: Prostaglandins and skin tumor promotion: Inhibition of tumor promotor-induced ornithine decarboxylase activity by inhibitors of prostaglandin synthesis. Biochem Biophys Res Commun (79): 1160-1166, 1977.
60. Verma AK, Ashendel CL, Boutwell RK: Inhibition by prostaglandin synthetase inhibitors of the induction of epidermal ornithine decarboxylase activity, the accumulation of prostaglandins and tumor promotion caused by 12-0-tetradecanoylphorbol-13-acetate. Cancer Res (40):308-315, 1980.
61. Furstenberger G, Marks F: Indomethacin inhibition of cell proliferation induced by phorbol ester TPA is reversed by prostaglandin E_2 in mouse epidermis in vivo. Biochem Biophys Res Commun (84):1103-1111, 1978.
62. Fischer SM, Gleason GL, Hardin LG, Bohrman JS, Slaga TJ: Prostaglandin modulation of phorbol ester skin tumor promotion. Carcinogenesis (1):245-248, 1980.
63. Levine L, Hassid A: Effects of phorbol-12,13-diesters on prostaglandin production and phospholipase activity in canine kidney (MDCK) cells. Biochem Biophys Res Commun (79):477-484, 1977.
64. Daniel LW, King L, Waite M: Source of arachidonic acid for prostaglandin synthesis in Madin-Darby canine kidney cells. J Biol Chem (256):12830-12835, 1981.
65. Tou JS: Activation of the metabolism of the fatty acyl group in granulocyte phospholipids by phorbol myristate acetate. Biochim Biophys Acta (665):491-497, 1981.
66. Ohuchi K, Levine L: Stimulation of prostaglandin synthesis by tumor promoting phorbol-12,13-diesters in canine kidney (MDCK) cells. J Biol Chem (253):4783-4790, 1978.
67. Ohuchi K, Levine L: α-Tocopherol inhibits 12-0-tetradecanoylphorbol-13-acetate-stimulated deacylation of cellular lipids, prostaglandin production, and changes in cell morphology of Madin-Darby canine kidney cells. Biochim Biophys Acta (619):11-19, 1980.
68. Mufson RA, DeFeo D, Weinstein IB: Effects of phorbol ester tumor promoters on arachidonic acid metabolism in

chick embryo fibroblasts. Molec Pharmacol (16):569-578, 1979.
69. Levine L, Ohuchi K: Retinoids as well as tumor promoters enhance deacylation of cellular lipids and prostaglandin production in MDCK cells. Nature (276):274-275, 1978.
70. Hirata F, Corcoran BA, Krishnamoorthy V, Schiffmann E, Axelrod J: Chemoattractants stimulate degradation of methylated phospholipids and release of arachidonic acid in rabbit leukocytes. Proc Natl Acad Sci USA (76): 2640-2643, 1979.
71. Bremer J, Figard PH, Greenberg DM: The biosynthesis of choline and its relation to phospholipid metabolism. Biochim Biophys Acta (43):477-488, 1960.
72. Sundler R, Akesson B: Regulation of phospholipid biosynthesis in isolated rat hepatocytes. J Biol Chem (250):3359-3367, 1975.
73. Hirata F, Axelrod J: Phospholipid methylation and biological signal transmission. Science (209):1082-1090, 1980.
74. Hirata F, Axelrod J: Enzymatic synthesis and rapid translocation of phosphatidylcholine by two methyltransferases in erythrocyte membrane. Proc Natl Acad Sci USA (75): 2348-2352, 1972.
75. Rovera G, Santoli D, Damsky C: Human promyelocytic leukemia cells in culture differentiate into macrophage-like cells when treated with a phorbol ester. Proc Natl Acad Sci USA (76):2779-2783, 1979.
76. Latern J, Sachs L: Regulation of normal differentiation in mouse and human myeloid leukemia cells by phorbol esters and the mechanism of tumor promotion. Proc Natl Acad Sci USA (76):5158-5162, 1979.
77. Cabot MC, Welsh CJ, Callahan MF, Huberman E: Alterations in lipid metabolism induced by 12-0-tetradecanoylphorbol-13-acetate in differentiating human myeloid leukemia cells. Cancer Res (40):3674-3679, 1980.
78. Hoffman DR, Huberman E: The control of phospholipid methylation by phorbol diesters in differentiating human myeloid HL-60 leukemia cells. Carcinogenesis (3):875-880, 1982.
79. Cassileth PA, Suholet D, Cooper RA: Early changes in phosphatidylcholine metabolism in human acute promyelocytic leukemia cells stimulated to differentiate by phorbol ester. Blood (58):237-243, 1981.

Chapter 5

TUMOR PROMOTING PHORBOL ESTERS MAY AFFECT CELL MEMBRANE SIGNAL
TRANSMISSION AND ARACHIDONATE METABOLISM BY MODULATING CALCIUM -
ACTIVATED, PHOSPHOLIPID - DEPENDENT PROTEIN KINASE

Curtis L. Ashendel

Purdue University

West Lafayette, IN 47907

		Page
I.	Phorbol ester receptor.	102
II.	Calcium-activated, phospholipid-dependent protein kinase and the phorbol ester receptor.	107
III.	Regulation of protein kinase c.	107
IV.	Role of PK-C in membrane signal transmission.	111
V.	Phorbol ester activation of PK-C.	114
VI.	Cell surface receptors for metabolites of arachidonic acid and phosphatidylinositol metabolism	117
VII.	PK-C and the regulation of arachidonic acid metabolism.	118
VIII.	Conclusions	118
References		119

S.M. Fischer and T.J. Slaga (eds.), ARACHIDONIC ACID METABOLISM AND TUMOR
PROMOTION. Copyright © 1985. Martinus Nijhoff Publishing, Boston. All rights reserved.

5. TUMOR PROMOTING PHORBOL ESTERS MAY AFFECT CELL MEMBRANE SIGNAL TRANSMISSION AND ARACHIDONATE METABOLISM BY MODULATING CALCIUM-ACTIVATED, PHOSPHOLIPID-DEPENDENT PROTEIN KINASE

Curtis L. Ashendel

Investigation of the molecular mechanism of action of phorbol esters has focused upon understanding how they promote tumors in mouse skin (1,2,3,4) and influence growth and differentiation of cells in culture (5,6,7,8). These studies have taken a major step forward with the identification and characterization of a receptor protein to which phorbol esters bind. These findings, especially aspects relevant to arachidonate metabolism, are reviewed below.

I. THE PHORBOL ESTER RECEPTOR

When phorbol esters were isolated from and characterized as the active principles of croton oil (9), it was postulated that they interacted specifically with cells (3). The relationship of the structure of phorbol esters to their irritancy and tumor promoting properties (10) as well as the very low effective doses of the most potent phorbol esters were consistent with a mechanism of action involving specific high affinity interaction with a receptor. A number of years elapsed before specific binding was observed because of problems due to the highly lipophilic character of the phorbol esters. However, since the first reports of specific binding to cell and tissue membranes of the tritium labeled phorbol esters [^{3}H]phorbol-12,13-dibutyrate ([^{3}H]PDBu) (11) and [^{3}H]12-0-tetradecanoylphorbol-13-acetate ([^{3}H]TPA) (12), research on the properties of the phorbol ester receptor has advanced briskly. Generally similar observations have been made using the [^{3}H]TPA and [^{3}H]PDBu binding assays, so it is often assumed that both assays measure binding to the same receptor.

The receptor displayed high affinity for reversibly binding [^{3}H]PDBu (dissociation equilibrium constant, K_d = 7 to 40 nM) (11,13,14,15,16) and [^{3}H]TPA (K_d = 66 to 1100 pM) (12,15,17). It specifically bound tumor promoting phorbol esters (11,16), mezerein (11,16), ingenol esters (16), teleocidin (18,19,20), and aplysiatoxin (20,21). It is interesting in light of the division of tumor promotion into multiple stages (22,23) that the second stage promoters, mezerein (22) and 12-O-retinoylphorbol-13-acetate (24,25,26,29), as well as the first stage promoter, 4-O-methyl-TPA (17,22), all bound to the receptor, although with lower affinity than TPA. Many other agents tested did not compete with phorbol esters for receptor binding, including antipromoting substances (11,14,16), prostaglandins (11,14,16), hormones (16), growth factors (11,14,16), neurotransmitters (27), and neuropharmacologic agents (27). A few compounds were able to compete for phorbol ester binding, but only with very low potency. These include arachidonic acid, the calcium ionophore A23187, phenothiazine antipsychotics (28), and quinidine (27).

Based upon binding affinity and competition of phorbol ester binding, some investigators have suggested that there are multiple distinct sites for binding these compounds (14,19,29,30,31). However, other data indicate only one class of binding sites (11,13,15,16,17,27,32,33,34,35,36,37,38,39) and biochemical evidence for distinct receptor proteins is lacking. Observations of distinct sites may merely have resulted from the same receptor protein in different environments (13,37,40), although the possibility of distinct receptor proteins has not been convincingly eliminated.

The phorbol ester receptor was initially identified as a protein since it was labile to treatment with heat, acid, and proteases (15,17). In addition, the binding activity in membrane preparations was partially inhibited by treatment with phospholipases, but not neuraminidase (15), suggesting that it was a non-glycosylated protein which was sensitive to its phospholipid environment. When the phorbol ester receptor was purified from mouse brain particulate protein, and was analyzed by sodium dodecylsulfate-polyacrylamide gel electrophoresis (SDS-PAGE), it was found to be a single polypeptide with an apparent M_r of 70,000 (41). When purified from rat brain soluble protein, the M_r of the receptor was 77,000 (42,43,44). Precise determination of the molecular weight will depend upon the development of antibody monospecific for the receptor.

The distribution of the phorbol ester receptor in cells, tissues, and organisms has provided the most useful clues to the biological function of this protein. Phorbol ester binding activity was reported to be in all vertebrate animals tested (45,47), as well as in the sea urchin (46), fruit fly (46), hydra (39), and nematode (47), but not in bacteria (11). It is not known whether the receptor or a similar protein is found in plant cells. These findings suggested that the phorbol ester receptor has an important biological function in animals. In the adult mouse, phorbol ester binding activity was most abundant in the brain (15,17,45) with more binding activity in the forebrain than the cerebellum which contains more than the spinal cord (27,45). Histological studies (48,49) have shown that the distribution of the receptor in the brain is distinct from the distribution of known neurotransmitters and their receptors. Although other tissues were shown to be rich in phorbol ester binding activity including spleen, thymus, and gastro-intestinal tract (15,17,45), the brain had between 2 and 4 times more activity than the tissue with the next highest level. The skin, adrenals, ovaries, fat, and salivary gland had less binding activity than the above tissues but more than the kidneys, liver, heart, or skeletal muscle. Although no receptor was detected in mouse erythrocytes, some was observed in chicken erythrocytes (50). Ontogenic studies, either autoradiography and histology of whole rat embryos (47) or disection of newborn mice (17,45) or chickens (27) revealed that the level of phorbol ester binding activity in the brain increases with development. These results indicated that the level of the phorbol ester receptor was not correlated with cell proliferation or metabolic activity. Taken together, the above distribution data suggested that the phorbol ester receptor may play a role in cell to cell communication, which is consistent with the observed effects of phorbol esters on cell to cell communication (51).

Since the phorbol ester receptor initially was not detected in the soluble cytosol of cells (36) or tissues (37), subcellular distribution studies focused on determination of the binding activity in the particulate fractions of cellular homogenates. These studies (37,52) did not conclusively demonstrate localization of the receptor to a single type of membrane, although the phorbol ester binding activity did not appear to be preferentially associated with nuclei or mitochondria. At best, it can be concluded that in homogenates, the receptor was localized in the lighter membrane fractions (endoplasmic reticulum and plasma membrane). During attempts to purify the receptor from soluble detergent extracts of mouse brain particulate protein, it was found that phorbol ester binding could only be measured in the presence of added phospholipid and calcium (41). When these conditions were employed it was possible to detect binding of phorbol esters by soluble cytoplastic

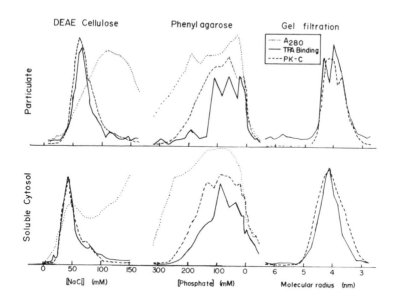

FIGURE 1. Identical chromatographic behavior of the phorbol ester receptor from the soluble or particulate fractions of mouse brain. Mouse brain soluble cytosol was prepared by extraction of mouse brains with 20 mM Tris buffer with 5 mM EDTA and DEAE cellulose chromatography included 5 mM EDTA in the elution buffer. The particulate fraction was washed with cold acetone and extracted with detergent and the DEAE cellulose was eluted with buffer containing detergent (41). Subsequent steps (hydroxylapatite, phenylagarose, and gel filtration were performed identically (41).

extracts of tissues (50,53,54,55) and cells (40). The soluble and membrane bound phorbol ester binding activities appeared to be the same protein which exists in two states - soluble and membrane-associated - which are in equilibrium, based upon the following evidence. The soluble and particulate proteins eluted identically from chromatographic columns (Figure 1), were both found in all tissues tested (53), and exhibited identical phorbol ester binding characteristics (53,55), suggesting that they are the same protein. That these exist in equilibrium was suggested by the observation that inclusion of EDTA in the homogenization buffer resulted in shifting the receptor to the soluble fraction while inclusion of calcium shifted it towards the particulate fraction (56). Figure 2 shows that calcium and TPA increased the binding to phospholipid of the phorbol ester receptor partially purified from rat brain cytosol. In fact, it has been hypothesized (57,58) that regulation of the biochemical function of the phorbol ester receptor is

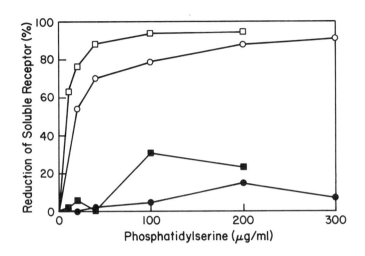

FIGURE 2. Association of the phorbol ester receptor with phosphatidylserine in the presence of TPA or calcium. The phorbol ester receptor purified from rat brain-cytosol on DEAE cellulose was mixed with the indicated concentration of phosphatidylserine and 1 mM EGTA (filled figures) or 0.3 mM $CaCl_2$ (open figures) with (squares) or without (circles) 24 nM TPA. After sedimentation at 48,000 x g for 30 min the specific [^{3}H]TPA binding activity in the soluble supernatent was determined (53).

accomplished by shifting the equilibrium between the membrane and soluble states of the receptor.

Despite accumulation of data on the biochemical and biological nature of the phorbol ester receptor, the assumption that it plays a role in tumor promotion or any other biological effect of phorbol esters has not been unequivocally demonstrated. One way to prove such a role would be to show that organisms or cells which did not have an active phorbol ester receptor did not respond to phorbol ester treatment, while those which otherwise were genetically identical did respond and had active receptor. This has not been successfully accomplished, although there have been many reports (30,31,33,36,59,60,61) of variant cell isolates which did not respond to phorbol esters but possessed normal numbers of receptors with usual binding activity. Particulate portions of epidermis of various strains of mice which differed in their sensitivity to tumor promotion bound phorbol ester indistinguishably (62). A reported (60) defect in receptor down-modulation in human promyelocytic cell variants which were not responsive to phorbol esters was not found to occur in other non-responsive cells (33,59) indicating that this was not a general phenomenon.

II. CALCIUM-ACTIVATED, PHOSPHOLIPID-DEPENDENT PROTEIN KINASE AND THE PHORBOL ESTER RECEPTOR

Partially purified calcium-activated, phospholipid-dependent protein kinase (PK-C) activity was shown to be increased by phorbol esters in the presence of suboptimal concentrations of calcium (56,63,64). This effect of phorbol esters in a cell free environment was unusual and it suggested that phorbol esters interacted directly with the protein kinase. This contention was supported by the partial copurification (56) of phorbol ester binding activity and PK-C activities from rat brain, as well as copurification to apparent homogenicity from mouse brain particulate protein (41) and rat brain cytosol (42,43,127). Although there were discrepancies in the apparent molecular weight of the receptor between these copurifications, the remainder of the characteristics described were in agreement. In addition, PK-C activity and the phorbol ester receptor have the same distributions in tissues, cells, and species (65,66).

It will be possible to provide better proof of identity of the phorbol ester receptor and PK-C when monospecific antibodies are produced. Until then, it remains as a working hypothesis that one protein is capable of both activities. Nevertheless, an accumulation of indirect pharmacologic data, consistent with the identity of the phorbol ester as PK-C are discussed in the next sections. In addition this hypothesis has significant appeal because it provides a ready explanation for the pleotypic effects of phorbol esters. Each of the cellular responses to phorbol esters could be due to phosphorylation of different proteins by PK-C. The observation that different cells respond differently might be the result of different availability of protein substrates or different effects of phosphorylation of the same proteins. In order to pursue beyond the above working hypothesis, the following discussion will examine what is known about this protein kinase and its regulation by endogenous and pharmacologic agents, what its cellular function may be, how this function is interfaced with other aspects of cellular regulation including the arachidonic acid cascade, and how these regulatory aspects are influenced by phorbol esters.

III. REGULATION OF PROTEIN KINASE C

A protein kinase which could be activated by limited proteolysis with trypsin (67) or either of two calcium-dependent proteases (68), was also found to become active in the presence of phospholipid and calcium (69). This protein kinase activity was independent of cyclic nucleotides or calmodulin and catalyticly unique (70,71,72,76) and so was considered to be distinct from other well characterized protein kinases. PK-C catalyzed transfer of phosphate to threonyl and seryl residues distinguishing it from tyrosine protein kinases (66,93,127). Although the proteolytic activation of PK-C yielded an enzyme fragment which was active without phospholipid or calcium (67,68) and proteolysis has severely hampered efforts purify the enzyme it remains to be determined whether this manner of regulating PK-C activity is significant in vivo. Interestingly, there have been two recent reports describing increased levels of the proteolyticly activated PK-C in platelets treated with phorbol esters or phospholipase C (73,74).

Without proteolysis, phospholipid or a hydrophobic substitute (75) always has been necessary in order to activate PK-C. Both the source and type of phospholipid influenced the amount of enzyme activation. Phosphatidylserine was found consistently to be the most effective while phosphatidylcholine always has been reported to be inactive or inhibitory for activation of PK-C (69,71,77). Other types of phospholipid, including sphingomyelin and phosphatidic acid were partially effective. Although phosphatidylserine partially activated PK-C in the presence of EGTA (65), the addition of calcium increased the stimulation of the enzyme. This effect was specific for calcium, as Mn, Cd, Co, Ni, and Ba were ineffective and Sr was only partially effective (69,71). Since activation of PK-C by phospholipid occured without calcium yet was enhanced by calcium, it is tempting to speculate that calcium enhanced the association of PK-C with phospholipid. The data shown in Figure 3 do not support this hypothesis since calcium did not reduce the amount of phospholipid required to maximally stimulate the kinase activity. This suggests that calcium did not increase the affinity of PK-C for phospholipid.

A number of agents other than phospholipid and calcium have been shown to affect PK-C. Exogenous compounds which inhibited PK-C such as phenothiazine anti-psychotic drugs (78), quercitin (79), polymyxin B (76), W-7 (76,80), adriamycin (85), alkyllysophospholipid (89) and mellitin (80) provide clues to the structural features of the enzyme which may be involved in its regulation. In addition to this, endogenous inhibitory compounds, such as retinal (81,82), polyamines (76,83), calmodulin (84), and acylcarnitines (76,85) also may function in the cell to regulate PK-C, although there currently is no evidence to support this. The inhibition of PK-C by palmitoyl carnitine, adriamycin, or trifluoperazine (85) was shown to be reduced at higher concentrations of

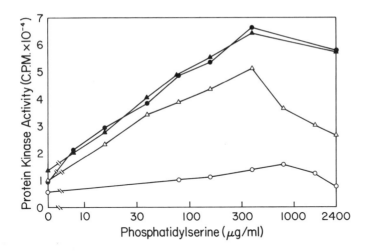

FIGURE 3. Activation of PK-C by TPA or calcium with various amounts of phosphatidylserine. Rat brain cytosolic phorbol ester receptor partially purified on DEAE-cellulose was assayed for PK-C activity (53) with the indicated amount of phosphatidylserine with 1 mM EGTA (open figures) or 0.01 mM $CaCl_2$ (solid figures) and 1.6% DMSO (circles) or 1.6 nM TPA and 1.6% DMSO (triangles).

calcium or phospholipid. Fluphenazine inhibition of PK-C was reduced by increased concentration of phospholipid (86). It was not clear whether this was due to competition for a similar binding site on the protein or due to dilution of the lipophilic inhibitors in the phospholipid bilayer.

Phorbol esters, certain other mouse skin tumor promoters, diacylglycerides, unsaturated fatty acids, fatty acylmonosaccharide lipid X, and retinoic acid can activate PK-C. More is known about the mechanisms of PK-C stimulation by phorbol esters and diacylglycerides than the latter three agents and these mechanisms are discussed at length in subsequent sections. The effect of unsaturated fatty acids (87) and retinoic acid (82) required added calcium but not phospholipid. Neither saturated fatty acids (87) nor other forms of vitamin A (81,82) activated PK-C. Although unesterified unsaturated fatty acids, as well as retinoic acid are present in animal cells, the concentrations in vivo are probably much below the 0.1 mM concentrations utilized in these studies. The concentration of unsaturated fatty acid used was approximately that which was shown to induce a respiratory burst in human neutrophils, an effect thought to be a result of PK-C activation (90). An important question is whether retinoic acid or unsaturated fatty acids play a

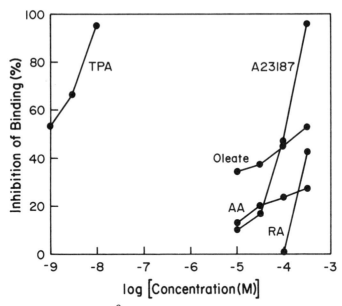

FIGURE 4. Inhibition of [^{3}H]TPA binding. The phorbol ester receptor was partially purified from rat brain cytosol by DEAE cellulose chromatography. TPA, oleate, archidonate (AA), retinoic acid (RA), and the calcium ionophore A23187 were included in the standard binding assay (53) at the concentrations indicated.

role in regulating PK-C in cells. At 0.1 mM, arachidonate reduced the K_a for calcium activation of PK-C in the presence of 20 micrograms/ml phosphatidylserine from 14.2 to 5.8 micromolar (87). The effects of lower concentrations of arachidonate or other unsaturated fatty acids on the activation of PK-C by calcium and phospholipid were not reported. It is interesting that high concentrations of retinoic acid, arachidonate, or oleate decreased phorbol ester binding to the receptor in the presence of phospholipid (Figure 4). Lipid X, N^2,O^3-diacylglucosamine-1-phosphate which is a precursor to bacterial lipid A, the biologically active lipopolysaccharide, stimulated PK-C in the presence of calcium but without added phosphatidylserine (88). It had also been shown that lipid X induced mitogenesis in B lymphocytes and activated macrophages, in a fashion similar to the effect of phorbol esters on those cells (7). Taken together the above data suggest that these acidic lipophilic compounds affect PK-C by the same mechanism as the acidic phospholipid phosphatidylserine.

When the calcium ionophore A23187 was used to treat cells or mouse skin it elicited some of the same responses as seen after treatment with phorbol esters (24). As can be seen in Figure 4, A23187 inhibited binding of [^{3}H]TPA to partially purified receptor, although the concentration required for this effect was higher than usually employed to affect cells, such as half-maximal stimulation of hexose transport in mouse fibroblasts by 1 μM of this compound (91). Furthermore, A23187 did not activate PK-C with or without calcium or phospholipid (data not shown). This suggested that this effect of A23187 on TPA binding, although direct, was not significant in comparison to its action as a calcium ionophore.

There have been three reports (92,93,127) of phosphorylation of PK-C in purified preparations, suggesting that it can phosphorylate itself. These reports depended upon proper identification of the M_r of PK-C and did not provide supporting evidence, such as a kinetic analysis and determination of the extent of phosphorylation. If PK-C was autophosphorylated, it would be interesting to determine the effect of this phosphorylation on the catalytic properties of this enzyme.

IV. THE ROLE OF PK-C IN MEMBRANE SIGNAL TRANSMISSION

When various fractions of membrane lipids were tested for their ability to activate PK-C, it was found that the neutral lipid fraction, while inactive by itself, could synergize with the phospholipid fraction for stimulation of the enzyme. Subsequent characterization showed that triacylglyceride, monoacylglyceride, and glycolipid did not further activate PK-C in the presence of suboptimal concentrations of phospholipid and calcium, but diacylglycerides (DAG) did (94,95). Comparisons of various synthetic DAG showed that the compound without at least one unsaturated fatty acid were almost inactive (94,95,96), and that potency increased with shortening the length of the fatty acyl group in 1-oleoyl-2-acylglyceride, as well as 1-acyl-2-oleoylglyceride (96). The two latter observations may have resulted from the poor solubility and slow micellular exchange of saturated DAG or DAG with two long chain fatty acids. Although not biologically relevant, it is interesting that 1-oleoyl-2-acetylglycerol was as active as 1-acetyl-2-oleoylglycerol (96), and also that 1,3-diolein was as effective as 1,2-diolein (95), indicating that orientation of the lipophilic fatty acid(s) relative to the free hydroxyl group did not affect the interaction with PK-C.

Although the data concerning the metabolism of phosphatidylinositol-diphosphate by phospholipase C to produce DAG in the membrane has been recently reviewed (58,98,99,100,101,102,103,104), it should be mentioned that there is little <u>direct</u> evidence for phospholipase C activation and increased DAG production in response to many hormones, neurotransmitters, and growth factors. Nevertheless, the concept of DAG as a second messenger for some growth factors, neurotransmitters and hormones is appealing because the ways in which many of these agents affect cells were shown to be similar to the effects of phorbol esters. Phorbol esters have been shown to affect the cell surface receptors for insulin (105), epidermal growth factor (106,107,108), transferrin (109,110), somatomedin C (105), and catecholamines (beta receptor) (111,112,113), as well as junctional cell to cell communication (114,115), suggesting that these agents affect some aspect of transmembrane signaling. A role for PK-C which is of fundamental importance to transmembrane signaling in cells also would be consistent with the phylogenic and tissue distribution of the phorbol ester receptor and also the inability to isolate viable mutant cells without functional phorbol ester receptors.

Data rapidly are accumulating which are consistent with a fundamental role for PK-C in transmembrane signaling and the identity of PK-C and the phorbol ester receptor, placing these hypotheses on a firm base of support. DAG inhibited phorbol ester binding to its receptor (116). In the following cases treatment of cells with phospholipase C or DAG mimicked treatment with phorbol esters: platelets were activated by treatment with phospholipase C (73) or DAG (117); while ornithine decarboxylase activity was induced in lymphocytes by treatment with phospholipase C (118); treatment of neutrophils with DAG induced the release of lysosomal enzymes (119,120) and the generation of superoxide (121); differentiation of human promyelocytic leukemia HL-60 cells to macrophages was induced by a DAG analog (122);and DAG produced the release of histamine from mast cells (130) and reduced the high affinity binding of EGF by 3T3 cells (107). In the reverse experiment, TPA induced contraction of isolated rabbit vascular smooth muscle an effect previously known to be induced by increased intracellular calcium, electrical stimulus or by treatment with alpha-adrenergic agonists (125). Alpha-adrenergic agonists have been thought to act via receptor-induced breakdown of phosphoinositides (104). Studies involving agents known to inhibit PK-C have found that these compunds blocked responses to TPA. Differentiation of HL-60 cells induced by TPA was inhibited by trifluoperazine (123), an inhibitor of PK-C. Promotion of mouse skin tumors by teleocidin was potently and totally inhibited by W-7

(124), another inhibitor of PK-C.

If PK-C does function in a way critically important to transmembrane signaling, that function has not been identified. It has been shown that DAG stimulated PK-C in vitro in the presence of suboptimal amounts of phospholipid and calcium and that DAG reduced the K_a for calcium (94,95). PK-C could be regulated in the same way in cells if the intracellular environment is insufficient to maximally activate PK-C. This enhancement of PK-C activation by calcium and phospholipid most likely occured by allosteric regulation. However, recalling that binding of phospholipid to PK-C was a prerequisite for its activation, the enhanced activation of PK-C by DAG must have been due either to DAG-enhanced association of PK-C with phospholipid or to some allosteric effect on PK-C distinct from and in addition to the effect of phospholipid and calcium. The appeal of the former hypothesis lies in the novelty of a mobile protein kinase, in which activation occurs with subcellular redistribution (57,58,126). Yet PK-C could partially be activated by phospholipid in the absence of calcium (Figure 3), suggesting that in cells, PK-C may be largely localized on membrane phospholipid. If this is the case and if DAG-activated PK-C is distinct from the calcium-activated form of the enzyme, then it is possible that the DAG-activated form is localized in regions of the membrane rich in DAG. Potential differences between the various activated forms of PK-C are discussed in the next section.

In addition to focusing on the mechanism of regulation of PK-C, recent research also has yielded some information concerning the substrate proteins for this enzyme. PK-C did not phosphorylate phosvitin or casein (67,76), but did phosphorylate basic proteins such as histones and protamine. All types of histones acted as substrates, but types H1 and H2B were found to be better substrates than the other types (69,76,77,128). Although phosphorylation of all substrates by PK-C was stimulated by phospholipid and calcium, phosphorylation of protamine was atypical in that it was much less dependent upon the presence of these activators (69,76,77,127). It should be noted that purified PK-C also phosphorylated histone in the absence of added calcium and phospholipid (65,71,77,94,95). While histones and protamine have been useful for assaying PK-C activity, their phosphorylation has not been considered biologically significant with respect to PK-C. In the search for biologically significant substrates for PK-C, the goal has been to explain how the activation of PK-C and resulting protein phosphorylation results in one or more of a variety of cellular responses. Although there are many cellular responses to agents which directly or indirectly activate PK-C, it is possible that this kinase may mediate the effects of these agents by phosphorylating one or a few proteins. This is not likely since the substrate specificity of PK-C is broad as indicated by the number of proteins for which there is some evidence for phosphorylation catalyzed by PK-C.

In many cases where phosphorylation of an isolated protein was shown to be catalyzed by calcium- and phospholipid-dependent kinase activity in a cell-free assay, it has not demonstrated that phosphorylation occured in cells in situations where PK-C would have been activated nor has the site of the phosphorylation been characterized. Proteins in these less than conclusive cases include retinoid-binding proteins (131), myelin basic protein (76), muscle phosphorylase kinase (72,76), phospholamban (132,133), glycogen synthetase (72,134), vinculin (135,136,137), filamin (137), th beta subunit of the eukaryotic initiation factor-2 (138), the alpha subunit of the inhibitory GTP binding protein (A. Gilman, personal communication), Polyoma virus middle T antigen (139), synapsin I, and the delta subunit of the acetylcholine receptor. Considerable evidence has accumulated that in intact cells, PK-C catalyzes the phosphorylation of a 40,000-47,000 M_r platelet protein (63,82,140), the light chain of myosin (141,142,143,144), the ribosomal S6 protein (43) and the receptor for epidermal growth factor (EGF) (93). The immediate effects of phosphorylation of the platelet protein remain unknown, while the phosphorylated EGF receptor displayed reduced tyrosyl kinase activity (93), the ATPase activity of phosphorylated myosin light chain was not activatable by actin, and phosphorylation of the ribosomal S6 protein has been correlated with increased protein synthesis. Despite these data, it is not clear how phosphorylation of any or all of these proteins could result in the various responses of cells to phorbol esters (2,5,6,7,8) or other agents thought to result in PK-C activation (58,97,99,101,103,104). Furthermore, it is difficult to understand why these phosphorylation events would be critical to the viability of animal cells.

V. PHORBOL ESTER ACTIVATION OF PK-C

The ability of phorbol esters to activate PK-C is the most recent finding of significance concerning the regulation of this enzyme. The first demonstration of the effect of these agents suggested that TPA mimicked the action of DAG on partially purified PK-C (63). In addition to reducing the calcium requirement for PK-C, TPA (64,145), DAG (145), and teleocidin (146) have been

Table 1. Catalytic properties of various active forms of PK-C.

Activators	K_M for histone H1 (μg/ml)	V_{Max} (pmol/min)
Phospholipid (20 μg/ml)	88	4.3
Phospholipid (8 μg/ml) + TPA (24 nM)	71	12.9
Phospholipid (8 μg/ml) + Calcium (0.3 mM)	3	7.2

shown to activate PK-C in the absence of calcium. Activation of PK-C by calcium or phorbol esters required phospholipid. The catalytic properties of calcium- and DAG-activated PK-C were compared with PK-C which was activated by TPA without calcium, and the K_M for histone H1 was found to be 4.5-fold higher for the TPA activated form (64). However, it is not clear whether the lower K_M was due to calcium or DAG. The data in Table 1 extend these findings and show that PK-C activated by TPA without calcium had a higher K_M for histone H1 than when activated with calcium, but somewhat lower than the K_M of the enzyme activated by phospholipid without calcium. These data suggest that calcium lowers the K_M for histone H1 whether or not TPA or DAG were present, and that the K_M for histone H1 is always high in the absence of calcium.

The effects of calcium, DAG, and TPA on PK-C have been studied by assaying phorbol ester binding activity rather than catalytic activity. Competitive inhibition of [^{3}H]PDBu binding by DAG has been observed (116), indicating interaction with PK-C at a common site. Calcium and phospholipid were required for maximal binding of [^{3}H]TPA to partially purified PK-C although [^{3}H]TPA did bind in the absence of calcium (Figure 5). It is interesting that inclusion of 1 mM $MgCl_2$ in the binding assay decreased the amount of calcium needed for maximal [^{3}H]TPA binding (Figure 5). Assays of PK-C activity usually included 5 to 10 mM magnesium which may explain why much more calcium was reported to be required for [^{3}H]TPA binding than for PK-C activation (40,41,42,64). Also shown in Figure 5 is that the K_a of calcium for enhancement of phorbol ester binding is reduced by increasing the phospholipid concentration. As determined by a quantitative sedimentation procedure, calcium and TPA affected the association of the receptor with phospholipid (Figure 2) in the same manner as these agents affected phospholipid activation of PK-C (Figure 3).

Although many effects of phorbol esters on cells have been reported, of particular relevance to this review is that phorbol esters induced a change in the apparent subcellular distribution of PK-C in cells (57,58,126). These observations that phorbol ester treatment of cells reduced levels of EDTA-

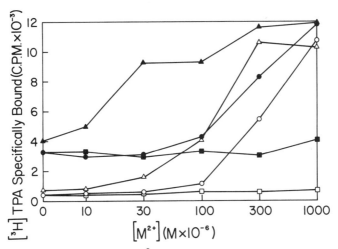

FIGURE 5. Dependence of [^{3}H]TPA binding calcium, magnesium, and phosphatidylserine. The 1 ml assays included the indicated concentrations of calcium (circles), magnesium (squares), or calcium with 1 mM magnesium (triangles) and either 1 microgram (open figures) or 25 micrograms (solid figures) of phosphatidylserine. The assay was done as described (53) using phorbol ester receptor purified from rat brain cytosol by DEAE-cellulose chromatography.

soluble PK-C and increased EDTA-insoluble PK-C probably resulted from the ability of phorbol esters to enhance the association of PK-C with phospholipid in the absence of calcium. Whether the apparent subcellular redistributions occurred in the intact cell or occurred as a result of the conditions used for cell disruption is not known.

The interest in whether DAG and TPA affected PK-C in the same manner stems from the differences in the observed effects on cells of phorbol esters and agents such as hormones and growth factors which increased DAG production in cells. There are at least four explanations for these differences. First, phorbol esters did not always increase intracellular calcium concentrations, while this was consistently observed with receptor-linked phosphoinositide metabolism. Another possibility is that interaction with DAG resulted in active PK-C which is catalyticly distinct from PK-C activated by phorbol esters. The localization of phorbol esters is likely to have been uniform throughout the plasma membrane and possibly in intracellular membranes. This could be quite different from the location of DAG which would be only in the plasma membrane, perhaps nonuniformly distributed. Lastly, the lifetime of DAG in cells was very short (63), while phorbol esters persisted much longer.

The similarities between the effects of phorbol esters and hormones and growth factors which are linked to phosphoinositide metabolism, as well as the differences are suggestive of further investigations of the regulation of PK-C. It seems clear, however, that these studies must be done with intact cells in order to directly relate the cell biology to the biochemical findings, and because PK-C is so sensitive to its ionic environment.

VI. CELL SURFACE RECEPTORS FOR METABOLITES OF ARACHIDONIC ACID AND PHOSPHATIDYLINOSITOL METABOLISM

Many cellular stimuli increase the metabolism of phosphatidylinositol (PI) leading to a transient increase in the levels of DAG in the cell membrane. However, binding of prostaglandin E (PGE) or prostacyclin (PGI$_2$) to the cell surface receptor for PGE did not lead to an increased PI metabolism (102,103,104,147,148), but rather to the activation of adenylate cyclase (102,103,104,148,169). The receptors for PGF have been shown to be distinct from the PGE receptor (150), but it is not known whether they are linked to PI metabolism or adenylate cyclase. There have been some reports linking breakdown of PI with binding of leukotriene B$_4$ (119) and thromboxanes (102,103,104) to cellular receptors. Distinct receptors for PGD and arachidonic acid hydroperoxide may exist but their mechanisms of transmembrane signaling have not been established.

PGs alone did not promote tumors in mouse skin, but modified the promotion of mouse skin tumors by TPA (151). The role of PGs and leukotrienes in tumor promotion is complicated (151) probably because of different effects of the various arachidonate metabolites (152). Studies using PG synthetase or lipoxygenase inhibitors suggested that at least some arachidonate metabolites were required for tumor promotion (151,153), inflammation (152), and some acute effects of phorbol esters on mouse skin (153,24) or cells in culture (154). Yet PGs alone were ineffective for eliciting these responses. Thus, it is not surprising that most PGs do not lead to the increased PI metabolism in cells. In general PGs have been considered as modifiers of the effects of other soluble regulators of cellular physiology (152).

Although mediation of the effects of PGs usually has been attributed to cyclic AMP (cAMP), increased intracellular calcium, either directly or indirectly, also has been implicated. An understanding of the molecular mechanism of regulation of intracellular calcium would allow determination of the role of calcium in mediating the effects of PGs on cells, and is the next major task confronting those interested in understanding the biochemistry of transmembrane signaling. Increased intracellular calcium might activate PK-C directly without increased production of DAG. Interestingly, the calcium ionophore A23187, which selectively increases intracellular calcium, induces some of the same cellular responses as caused by PGs and phorbol esters. These include: T lymphocyte mitogenesis (139); the first stage of mouse skin tumor promotion (155); activation of neutrophils (119); and mitogenesis in mouse skin (24). In many instances A23187 synergized with phorbol esters or endogenous growth regulatory substances for their effects on cell physiology. Of course, PGs have effects on cells other than increasing intracellular calcium or cAMP concentrations, and calcium affects many cellular constituents in addition to PK-C. The knowledge that phorbol esters directly activate PK-C should make easier the task of unraveling the roles of calcium, PK-C, and cAMP in mediating the cellular responses to PGs.

VII. PK-C AND THE REGULATION OF ARACHIDONIC ACID METABOLISM

Treatment of many types of cells (156) or mouse skin (24,157) with phorbol esters resulted in increased levels of PGs. The rate limiting step of PG biosynthesis is the release of free fatty acids from phospholipids catalyzed by phospholipase A_2 or from diacylglycerol catalyzed by diacylglyceride lipase (158). This enzyme is regulated by an inhibitory polypeptide (lipomodulin) which could be induced by glucocorticosteroids (158). Phosphorylation of lipomodulin has also been reported to decrease its inhibition of phospholipase A_2 (161). One mechanism by which phorbol esters might activate phospholipase A_2 is by activation of PK-C resulting in activation of a tyrosine kinse leading to phosphorylation of lipomodulin (139). Indirectly increasing intracellular calcium would be another way phorbol esters could stimulate PG biosynthesis, since the calcium ionophore A23187 increased PGE levels in mouse skin (24).

VIII. CONCLUSIONS

It has been well documented (151) that prostaglandins or other metabolites of arachidonic acid are essential for tumor promotion in mouse skin, particularly the second stage of the promotion. Equally well documented is the role of PGs in tissue inflammation (152), including mouse skin treated with phorbol esters (154). Our understanding of how phorbol esters affect PG synthesis has progress rapidly with the identification and characterization of the phorbol ester receptor as PK-C, as well as studies of the substrates for and regulation of PK-C. That PK-C was activated by phorbol esters, many growth factors, and increased concentrations of intracellular calcium provided explanations for many of the effects of phorbol esters and the calcium ionophore A23187. It seems that an explanation is close at hand for the mechanism of phorbol ester-induced inflammation which is an essential component of tumor promotion in mouse skin.

These findings also have suggested explanations for the non-inflammatory components of mouse skin tumor promotion. These somewhat obscure components are likely to include a special kind of hyperplasia involving an increased number of dark cells (159,160). Also they are sufficient for the first stage of mouse skin tumor promotion (19,155) which was sensitive to inhibition by glucocorticosteroids and protease inhibitors (19). Phorbol esters may have induced these effects in mouse skin by mimicking or enhancing the action of a growth factor as was observed in vitro, by antagonizing the action of a growth inhibitory factor such as a chalone, or by altering cellular sensing of neighboring cells. In each case, phorbol esters bind to and activate PK-C and alter cellular transmembrane signaling resulting in cellular misperception of the extracellular environment. In other words, the cells treated with phorbol esters hallucinate and respond in ways which separately are normal, but which occur in bizarre and unnatural combinations.

ACKNOWLEGEMENTS

The technical assistance of P. Minor, L. Hermsen, and B. Stenko was greatly appreciated. This research was supported by American Cancer Society institutional research grant IN-17V-5 and NIH research grants number CA 36262 and CA 23168. The author is a recipient of Junior Faculty Research Award number 106 from the American Cancer Society.

REFERENCES
1. Boutwell RK, Verma AK, Ashendel CL, Astrup E: Mouse skin: A useful model system for studying the mechanism of chemical carcinogenesis. In: Hecker E, Fusenig NE, Kunz W, Marks F, Thielmann HW (eds) Carcinogenesis a comprehensive survey, Volume 7: Cocarcinogenesis and biological effects of tumor promoters. Raven Press, New York, 1982, pp 1-12.
2. Boutwell RK: The function and mechanism of promoters of carcinogenesis. CRC Crit Rev Toxicol (2): 419-443, 1974.
3. Van Duuren BL: Tumor-promoting agents in two-stage carcinogenesis. Progr Exp Tumor Res (11): 31-68, 1969.
4. Diamond L, O'Brien TG, Baird W: Tumor promoters and the mechanism of tumor promotion. Adv Cancer Res (32): 1-74, 1980.
5. Blumberg PM: In vitro studies on the mode of action of the phorbol esters, potent tumor promoters. CRC Crit Rev Toxicol (8): 153-234, 1980.

6. Diamond L, O'Brien TG, Rovera G: Tumor promotors: Effects on proliferation and differentiation of cells in culture. Life Sci (23): 1979-1988, 1978.
7. Mastro AM: Phorbol esters: Tumor promotion, cell regulation, and the immune response. Lymphokines (6): 263-313, 1982.
8. Yuspa SH, Hennings H, Lichti U: Initiator and promoter induced specific changes in epidermal function and biological potential. J Supramol Struct Cell Biochem (17): 245-257, 1981.
9. Van Duuren BL, Orris L: The tumor-enhancing principles of Croton tiglum L. Cancer Res (25): 1871-1875, 1965.
10. Hecker E: Structure-activity relationships in diterpene esters irritant and cocarcinogenic to mouse skin. In: Slaga TJ, Sivak A, Boutwell RK (eds) Carcinogenesis a comprehensive survey, Volume 2: Mechanisms of tumor promotion and cocarcinogenesis. Raven Press, New York, 1978, pp 11-48.
11. Driedger PE, Blumberg PM: Specific binding of phorbol ester tumor promoters. Proc Natl Acad Sci USA (77): 567-571, 1980.
12. Ashendel CL, Boutwell RK: Direct measurement of specific binding of highly lipophilic phorbol diester to mouse epidermal membranes using cold acetone. Biochem Biophys Res Commun (99): 543-549, 1981.
13. Jaken S, Feldman H, Blumberg PM, Tashjian AH: Association of phorbol ester receptor down modulation with a cryptic receptor state. Cancer Res (43): 5795-5800, 1983.
14. Delclos KB, Nagle DS, Blumberg PM: Specific binding of phorbol ester tumor promoters to mouse skin. Cell (19): 1025-1032, 1980.
15. Dunphy WG, Delclos KB, Blumberg PM: Characterization of specific binding of [^{3}H]phorbol 12,13-dibutyrate and [^{3}H]phorbol 12-myristate 13-acetate to mouse brain. Cancer Res (40): 3635-3641, 1980.
16. Shoyab M, Todaro GJ: Specific high affinity cell membrane receptors for biologically active phorbol and ingenol esters. Nature (288): 451-455, 1980.
17. Hergenhahn M, Hecker E: Specific binding of the tumor promoter TPA in various mouse organs as measured by a "cold acetone-filter assay." Carcinogenesis (2): 1277-1281, 1981.
18. Umezawa K, Weinstein IB, Horowitz A, Fujiki H, Matsushima T, Sugimura T: Similarity of teleocidin B and phorbol ester tumor promoters in effects on membrane receptors. Nature (290): 411-413, 1981.
19. Horowitz AD, Greenebaum E, Weinstein IB: Identification of receptors for phorbol ester tumor promoters in intact mammalian cells and of an inhibitor of receptor binding in biologic fluids. Proc Natl Acad Sci USA (78): 2315-2319, 1981.
20. Horowitz AD, Fujiki H, Weinstein IB, Jeffrey A, Okin E, Moore RE, Sugimura T: Comparative effects of aplysiatoxin, debromoaplysiatoxin, and teleocidin on receptor binding and phospholipid metabolism. Cancer Res (43): 1529-1535, 1983.
21. Suganuma M, Fujiki H, Tahira T, Cheuk C, Moore RE, Sugimura T: Estimation of tumor promoting activity and structure-function relationships of aplysiatoxins. Carcinogenesis (5): 315-318, 1984.
22. Slaga TJ, Fischer SM, Nelson K, Gleason GL: Studies on the mechanism of skin tumor promotion: Evidence for several stages in promotion. Proc Natl Acad Sci USA (77): 3659-3663, 1980.
23. Slaga TJ, Fischer SM, Weeks CE, Nelson K, Mamrack M, Klein-Szanto AJP: Specificity and mechanism(s) of promoter inhibitors in multistage promotion. In: Hecker E, Fusenig NE, Kunz W, Marks F, Thielmann HW (eds) Carcinogenesis a comprehensive survey, Volume 7: Cocarcinogenesis and

biological effects of tumor promoters. Raven Press, New York, 1982, pp 19-34.
24. Marks F, Berry DL, Bertsch S, Furstenberger G, Richter H: On the relationship between epidermal hyperproliferation and skin tumor promotion. In: Hecker E, Fusenig NE, Kunz W, Marks F, Thielmann HW (eds) Carcinogenesis a comprehensive survey, Volume 7: Cocarcinogenesis and biological effects of tumor promoters. Raven Press, New York, 1982, pp 331-346.
25. Furstenberger G, Berry DL, Sorg B, Marks F: Skin tumor promotion by phorbol esters in a two-stage process. Proc Natl Acad Sci USA (78): 7722-7726, 1981.
26. Berry DL: Structure-activity relationships with a series of synthetic phorbol diesters in mouse epidermal cells in culture. Proc Am Assoc Cancer Res (23): 102, 1982.
27. Nagle DS, Jaken S, Castagna M, Blumberg PM: Variation with embryonic development and regional localization of specific [^{3}H]phorbol 12,13-dibutyrate binding to brain. Cancer Res (41): 89-93, 1981.
28. Shoyab M, Todaro GJ, Tallman JF: Chlorpromazine and related antipsychotic tricyclic compounds competitively inhibit the interaction between tumor-promoting phorbol esters and their specific receptors. Cancer Lett (16): 171-177, 1982.
29. Dunn JA, Blumberg PM: Specific binding of [20-^{3}H]12-deoxyphorbol 13-isobutyrate to phorbol ester receptor subclasses in mouse skin particulate preparations. Cancer Res (43): 4632-4637, 1983.
30. Colburn NH, Gindhart TD, Dalal B, Hegamyer GA: The role of phorbol ester receptor binding in responses to promoters by mouse and human cells. In: Langenbach R, Nesnow S, Rice JM (eds) Organ and species specificity in chemical carcinogenesis. Plenum Publishing Corp, New York, 1983, pp 189-200.
31. Spinelli W, Ishii DN: Tumor promoter receptors regulating neurite formation in cultured human neuroblastoma cells. Cancer Res (43): 4119-4125, 1983.
32. Sando JJ, Hilfiker ML, Salomon DS, Farrar JJ: Specific receptors for phorbol esters in lymphoid cell populations: Role in enhanced production of T-cell growth factor. Proc Natl Acad Sci USA (78): 1189-1193, 1981.
33. Lehrer RI, Cohen LE, Koeffler HP: Specific binding of [^{3}H]phorbol dibutyrate to phorbol diester-responsive and -resistant clones of a human myeloid leukemia (KG-1) line 1. Cancer Res (43): 3563-3566, 1983.
34. Jones MJ, Murray AW: Decreased [20-^{3}H]phorbol-12,13-dibutyrate binding to phospholipase C pretreated epidermal cells. Cancer Lett (19): 91-98, 1983.
35. Cooper RA, Braunwald AD, Kuo AL: Phorbol ester induction of leukemic cell differentiation is a membrane-mediated process. Proc Natl Acad Sci USA (79): 2865-2869, 1982.
36. Blumberg PM, Butler-Gralla E, Herschman HR: Analysis of phorbol ester receptors in phorbol ester unresponsive 3T3 cell variants. Biochem Biophys Res Commun (102): 818-823, 1981.
37. Dunphy WG, Kochenburger RJ, Castagna M, Blumberg PM: Kinetics and subcellular localization of specific [^{3}H]phorbol 12,13-dibutyrate binding by mouse brain. Cancer Res (41): 2640-2647, 1981.
38. Chida K, Kuroki T: Presence of specific binding sites for phorbol ester tumor promoters in human epidermal and dermal cells in culture but lack of down regulation in epidermal cells. Cancer Res (43): 3638-3642, 1983.
39. Yamasaki H, Drevon C, Martel N: Specific binding of phorbol esters to Friend erythroleukemia cells - general properties, down regulation and relationship to cell differentiation. Carcinogenesis (3): 905-910, 1982.

40. Sando JJ, Young MC: Identification of high-affinity phorbol ester receptor in cytosol of EL4 thymoma cells: Requirement for calcium, magnesium, and phospholipids. Proc Natl Acad Sci USA (80): 2642-2646, 1983.
41. Ashendel CL, Staller, JM, Boutwell RK: Solubilization, purification, and reconstitution of a phorbol ester receptor from the particulate protein fraction of mouse brain. Cancer Res (43): 4327-4332, 1983.
42. Kikkawa U, Takai Y, Tanaka Y, Miyake R, Nishizuka Y: Protein kinase C as a possible receptor protein of tumor-promoting phorbol esters. J Biol Chem (258): 11442-11445, 1983.
43. LePeuch CJ, Ballester R, Rosen OM: Purified rat brain calcium- and phospholipid- dependent protein kinase phosphorylates ribosomal protein S6. Proc Natl Acad Sci USA (80): 6858-6862, 1983.
44. Wolf M, Sahyoun N, LeVine H, Cuatrecasas P: Protein kinase C: Rapid enzyme purification and substrate-dependence of the diacylglycerol effect. Biochem Biophys Res Commun (122): 1268-1275, 1984.
45. Shoyab M, Warren TC, Todaro GJ: Tissue and species distribution and developmental variation of specific receptors for biologically active phorbol and ingenol esters. Carcinogenesis (2): 1273-1276, 1981.
46. Blumberg PM, Delclos KB, Jaken S: Tissue and species specificity for phorbol ester receptors. In: Langenbach R, Nesnow N, and Rice JM (eds) Organ and species specificity in chemical carcinogenesis. Plenum Publishing Corp, New York, 1983, pp 201-229.
47. Lew KK, Chritton S, Blumberg PM: Biological responsiveness to phorbol esters and specific binding of [^{3}H]phorbol 12,13-dibutyrate in the nematode Caenorhabditis elegans, a manipulable genetic system. Teratogen Carcinogen Mutagen (2): 19-30, 1982.
48. Nagle DS, Blumberg PM: Regional localization by light microscopic autoradiography of receptors in mouse brain for phorbol ester tumor promoters. Cancer Lett (18): 35-40, 1983.
49. Murphy KMM, Gould RJ, Oster-Granite ML, Gearhart JD, Snyder SH: Phorbol ester receptors: Autoradiographic identification in the developing rat. Science (222): 1036-1038, 1983.
50. Gschwendt M, Kittstein W: Soluble phorbol ester binding sites in cytosol of chick liver and brain - Activation by phosphatidylserine. Cancer Lett (19), 159-164, 1983.
51. Trosko JE, Yotti LP, Warren ST, Tsushimoto G, Chang C: Inhibition of cell-cell communication by tumor promoters. In: Hecker E, Fusenig NE, Kinz W, Marks F, Thielmann HW (eds) Carcinogenesis a comprehensive survey, Volume 7: Cocarcinogenesis and biological effects of tumor promoters. Raven Press, New York, 1982, pp 565-585.
52. Blumberg PM, Delclos KB, Dunphy WG, Jaken S: Specific binding of phorbol ester tumor promoters to mouse tissues and cultured cells. In: Hecker E, Fusenig NE, Kunz W, Marks F, Thielmann HW (eds) Carcinogenesis a comprehensive survey, Volume 7: Cocarcinogenesis and biological effects of tumor promoters. Raven Press, New York, 1982, pp 519-536.
53. Ashendel CL, Staller JM, Boutwell RK: Identification of a calcium- and phospholipid-dependent phorbol ester binding activity in the soluble fraction of mouse tissues. Biochem Biophys Res Commun (111): 340-345, 1983.
54. Gschwendt M, Horn F, Kittstein W, Furstenberger G, Marks F: Soluble phorbol ester binding sites and phospholipid- and calcium-dependent protein kinase activity in cytosol of chick oviduct. FEBS Lett (162): 147-150, 1983.
55. Leach KL, James ML, Blumberg PM: Characterization of a specific phorbol ester aporeceptor in mouse brian cytosol. Proc Natl Acad Sci USA (80):

4208-4212, 1983.
56. Niedel JE, Kuhn LJ, Vandenbark GR: Phorbol diester receptor copurifies with protein kinase C. Proc Natl Acad Sci USA (80): 36-40, 1983.
57. Kraft AD, Anderson WB: Phorbol esters increase the amount of Ca^{2+}, phospholipid-dependent protein kinase associated with plasma membrane. Nature (301): 621-623, 1983.
58. Nishizuka Y: The role of protein kinase C in cell surface signal transduction and tumor promotion. Nature (308): 693-698, 1984.
59. Weinberg JB, Misukonis MA, Goodwin BJ: Human Leukemia Cell Lines with comparable receptor binding characteristics but different phenotypic responses to phorbol diesters. Cancer Res (44): 976-980, 1984.
60. Solanki V, Slaga TJ, Callaham M, Huberman E: Down regulation of specific binding of [20-^{3}H]phorbol 12,13-dibutyrate and phorbol ester-induced differentiation of human promyelocytic leukemia cells. Proc Natl Acad Sci USA (78): 1722-1725, 1981.
61. Sando JJ, Hilfiker ML, Piacentini MJ, Laufer TM: Identification of phorbol ester receptors in T-cell growth factor-producing and -nonproducing EL4 mouse thymoma cells. Cancer Res (42): 1676-1680, 1982.
62. Wheldrake JF, Marshall J, Ramli J, Murray AW: Skin carcinogenesis and promoter binding characteristics in different mouse strains. Carcinogenesis (3): 805-807, 1982.
63. Castagna M, Takai Y, Kaibuchi K, Sano K, Kikkawa U, Nishizuka Y: Direct activation of calcium-activated, phospholipid-dependent protein kinase by tumor-promoting phorbol esters. J Biol Chem (257), 7847-7851, 1982.
64. Vandenbark GR, Kuhn LJ, Niedel JE: Possible mechanism of phorbol diester-induced maturation of human promyelocytic leukemia cells. (73): 48-457, 1984.
65. Kuo JF, Andersson RGG, Wise BC, Mackerlova L, Salomonsson I, Brackett NL, Katoh N, Shoji M, Wrenn RW: Calcium-dependent protein kinase: Widespread occurrence in various tissues and phyla of the animal kingdom and comparison of effects of phospholipid, calmodulin, and trifluoperazine. Proc Natl Acad Sci USA (77): 7039-7043, 1980.
66. Minakuchi R, Takai Y, Yu B, Nishizuka Y: Widespread occurrence of calcium-activated, phospholipid-dependent protein kinase in mammalian tissues. J Biochem (89): 1651-1654, 1981.
67. Inoue M, Kishimoto A, Takai Y, Nishizuka Y: Studies on a cyclic nucleotide-independent protein kinase and its proenzyme in mammalian tissues. J Biol Chem (252): 7610-7616, 1977.
68. Kishimoto A, Kajikawa N, Shiota M, Nishizuka Y: Proteolytic activation of calcium-activated, phospholipid-dependent protein kinase by calcium-dependent neutral protease. J Biol Chem (258): 1156-1164, 1983.
69. Takai Y, Kishimoto A, Iwasa Y, Kawahara Y, Mori T, Nishizuka Y: Calcium-dependent activation of a multifunctional protein kinase by membrane phospholipids. J Biol Chem (254): 3692-3695, 1979.
70. Takai Y, Kishimoto A, Inoue M, Nishizuka Y: Studies on a cyclic nucleotide-independent protein kinase and its proenzyme in mammalian tissues. J Biol Chem (252): 7603-7609, 1977.
71. Wise BC, Raynor RL, Kuo JF: Phospholipid-sensitive Ca^{2+}-dependent protein kinase from heart. J Biol Chem (257): 8481-8488, 1982.
72. Nishizuka Y, Takai Y, Hashimoto E, Kishimoto A, Kuroda Y, Sakai K, Yamamura H: Regulatory and functinal compartment of three multifunctinal protein kinase systems. Mol Cell Biochem (23): 153-165, 1979.
73. Tapley PM, Murray AW: Platelet Ca^{2+}-activated, phospholipid-dependent protein kinase: Evidence for proteolytic activation of the enzyme in cells treated with phospholipase C. Biochem Biophys Res Commun (118):

835-841, 1984.
74. Tapley PM, Murray AW: Modulation of Ca^{2+}-activated, phospholipid-dependent protein kinase in platelets treated with a tumor-promoting phorbol ester. Biochem Biophys Res Commun (122): 158-164, 1984.
75. Zabrenetzky VS, Bruckwick E, Lovenberg W: Calcium stimulation of protein kinase C in the absence of added phospholipids. Biochem Biophys Res Commun (102): 135-141, 1981.
76. Wise BC, Glass DB, Chou C-HJ, Raynor RL, Katoh N, Schatzman RC, Turner RS, Kibler RF, Kuo JF: Phospholipid-sensitive Ca^{2+}-dependent protein kinase from heart. J Biol Chem (257): 8489-8495, 1982.
77. Ashendel CL, Staller JM, Boutwell, RK: Protein kinase activity associated with a phorbol ester receptor purified from mouse brain. Cancer Res (43): 4333-4337, 1983.
78. Wrenn RW, Katoh N, Schatzman RC, Kuo JF: Inhibition by phenothiazine antipsychotic drugs of calcium-dependent phosphorylation of cerebral cortex proteins regulated by phospholipid or calmodulin. Life Sci (29): 725-733, 1981.
79. Gschwendt M, Horn F, Kittstein W, Marks F: Inhibition of the calcium- and phospholipid-dependent protein kinase activity from mouse brain cytosol by quercetin. Biochem Biophys Res Commun (117): 444-447, 1983.
80. Helfman DM, Appelbaum BD, Vogler WR, Kuo JF: Phospholipid-sensitive Ca^{2+}-dependent protein kinase and its substrates in human neutrophils. Biochem Biophys Res Commun (111): 847-853, 1983.
81. Taffet SM, Greenfield ARL, Haddox MK: Retinal inhibits TPA activated, calcium-dependent, phospholipid-dependent protein kinase ("C" kinase). Biochem Biophys Res Commun (114): 1194-1199, 1983.
82. Ohkubo S, Yamada E, Endo T, Itoh H, Hidaka H: Vitamin A acid-induced activation of Ca^{2+}-activated, phospholipid-dependent protein kinase from rabbit retina. Biochem Biophys Res Commun (118): 460-466, 1984.
83. Qi D, Shatzman RC, Mazzei GJ, Turner RS, Raynor RL, Liao S, Kuo JF: Polyamines inhibit phospholipid-sensitive and calmodulin-sensitive Ca^{2+}-dependent protein kinases. Biochem J (213): 281-288, 1983.
84. Albert KA, Wu WC-S, Nairn AC, Greengard P: Inhibition by calmodulin of calcium/phospholipid-dependent protein phosphorylation. Proc Natl Acad Sci USA (81): 3622-3625, 1984.
85. Wise BC, Kuo JF: Modes of inhibition by acylcarnitines, adriamycin and trifluoperazine of cardiac phospholipid-sensitive calcium-dependent protein kinase. Biochem Pharmacol (32): 1259-1265, 1983.
86. Donnelly TE, Jensen R: Effect of fluphenazine on the stimulation of calcium-sensitive phospholipid-dependent protein kinase by 12-0-tetradecanoyl phorbol-13-acetate. Life Sci (33): 2247-2253, 1983.
87. McPhail LC, Clayton CC, Snyderman R: A potential second messenger role for unsaturated fatty acids: Activation of Ca^{2+}-dependent protein kinase. Science (224): 622-625, 1984.
88. Wightman PD, Raetz CRH: The activation of protein kinase C by biologically active lipid moieties of lipopolysaccharide. J Biol Chem (259): 10048-10052, 1984.
89. Helfman DM, Barnes KC, Kinkade JM, Vogler WR, Shoji M, Kuo JF: Phospholipid-sensitive Ca^{2+}-dependent protein phosphorylation system in various types of leukemic cells from human patients and in human leukemic cell lines HL60 and K562, and its inhibition by alkyllysophospholipid. Cancer Res (43): 2655-2961, 1983.
90. Badwey JA, Curnutte JT, Karnovsky ML: cis-Polyunsaturated fatty acids induce high levels of superoxide production of human neutrophils. J Biol Chem (256): 12640-12643, 1981.

91. Yamanishi K, Nishino H, Iwashima A: Ca^{2+}-dependent stimulation of hexose transport by A23187, 12-O-tetradecanoylphorbol-13-acetate and epidermal growth factor in mouse fibroblasts. Biochem Biophys Res Commun (117): 637-642, 1983.
92. Sano K, Takai Y, Yamanaishi J, Nishizuka J: A role of calcium-activated phospholipid-dependent protein kinase in human platelet activation. J Biol Chem (258): 2010-2013, 1983.
93. Cochet C, Gill GN, Meisenhelder J, Cooper JA, Hunter T: C-Kinase phosphorylates the epidermal growth factor receptor and reduces its epidermal growth factor-stimulated tyrosine protein kinase activity. J Biol Chem (259): 2553-2558, 1984.
94. Takai Y, Kishimoto A, Kikkawa U, Mori T, Nishizuka Y: Unsaturated diacylglycerol as a possible messenger for the activation of calcium-activated, phospholipid-dependent protein kinase system. Biochem Biophys Res Commun (91): 1218-1224, 1979.
95. Kishimoto A, Takai Y, Mori T, Kikkawa U, Nishizuka Y: Activation of calcium and phospholipid-dependent protein kinase by diacylglycerol, its possible relation to phosphatidylinositol turnover. J Biol Chem (255): 2273-2276, 1980.
96. Mori T, Takai Y, Yu B, Takahashi J, Nishizuka Y, Fujikura T: Specificity of the fatty acyl moieties of diacylglycerol for the activation of calcium-activated, phospholipid-dependent protein kinase. J Biochem (91): 427-431, 1982.
97. Gil DW, Brown SA, Seeholzer SH, Wildey GM: Phosphatidylinositol turnover and cellular function. Life Sci (32): 2034-2046, 1983.
98. Agranoff BW: Biochemical mechanisms in the phosphatidylinositol effect. Life Sci (32): 2047-2054, 1983.
99. Fain JN, Lin SH, Litosch I, Wallace M: Hormonal regulation of phosphatidylinositol breakdown. Life Sci (32): 2055-2067, 1983.
100. Lapetina EG: Metabolism of inositides and the activation of platelets. Life Sci (32): 2069-2082, 1983.
101. Michell RH: Polyphosphoinositide breakdown as the initiating reaction in receptor-stimulated inositol phospholipid metabolism. Life Sci (32): 2083-2085, 1983.
102. Lapetina EG, Siess W: The role of phospholipase C in platelet responses. Life Sci (33): 1011-1018, 1983.
103. Nishizuka Y: Calcium, phospholipid turnover and transmembrane signalling. Phil Trans R Soc Lond B (302): 101-112, 1983.
104. Nishizuka Y: Turnover of inositol phospholipids and signal transduction. Science (225): 1365-1370, 1984.
105. Jacobs S, Sahyoun NE, Saltiel AR, Cuatrecasas P: Phorbol esters stimulate the phosphorylation of receptors for insulin and somatomedin C. Proc Natl Acad Sci USA (80): 6211-6213, 1983.
106. Davis RJ, Czech MP: Tumor-promoting phorbol diesters mediate phosphorylation of the epidermal growth factor receptor. J Biol Chem (259): 8545-8549, 1984.
107. Brown KD, Blay J, Irvine RF, Heslop JP, Berridge MJ: Reduction of epidermal growth factor receptor affinity by heterologous ligands: Evidence for a mechanism involving the breakdown of phosphoinositides and the activation of protein kinase C. Biochem Biophys Res Commun (123): 377-384, 1984.
108. Moon SO, Palfrey HC, King AC: Phorbol esters potentiate tyrosine phosphorylation of epidermal growth factor receptors in A431 membranes by a calcium-independent mechanism. Proc Natl Acad Sci USA (81): 2298-2302, 1984.

109. May WS, Jacobs S, Cuatrecasas P: Association of phorbol ester-induced hyperphosphorylation and reversible regulation of transferrin membrane receptors in HL60 cells. Proc Natl Acad Sci USA (81): 2016-2020, 1984.
110. Klausner RD, Harford J, Van Renswoude J: Rapid internalization of the transferrin receptor in K562 cells is triggered by ligand binding or treatment with a phorbol ester. Proc Natl Acad Sci USA (81): 3005-3009, 1984.
111. Garte SJ, Currie D, Belman S: Inhibition of β-adrenergic response in cultured epidermal cells by phorbol myristate acetate. Carcinogenesis (4): 939-940, 1983.
112. Sibley DR, Nambi P, Peters JR, Lefkowitz RJ: Phorbol diesters promote β-adrenergic receptor phosphorylation and adenylate cyclase desensitization in duck erythrocytes. Biochem Biophys Res Commun (121): 973-979, 1984.
113. Kelleher DJ, Pessin JE, Ruoho AE, Johnson GL: Phorbol ester induces desensitization of adenylate cyclase and phosphorylation of the β-adrenergic receptor in turkey erythrocytes. Proc Natl Acad Sci USA (81): 4316-4320, 1984.
114. Yamasaki H, Enomoto T, Martel N, Shiba Y, Kanno Y: Tumor promoter-mediated reversible inhibition of cell-cell communication (electrical coupling). Exp Cell Res (146): 297-308, 1983.
115. Fitzgerald DJ, Knowles SE, Ballard FJ, Murray AW: Rapid and reversible inhibition of junctional communicatin by tumor promoters in a mouse cell line. Cancer Res (43): 3614-3618, 1983.
116. Sharkey NA, Leach KL, Blumberg PM: Competitive inhibition by diacylglycerol of specific phorbol ester binding. Proc Natl Acad Sci USA (81): 607-610, 1984.
117. Kaibuchi K, Sano K, Hoshijima M, Takai Y, Nishizuka Y: Phosphatidyl-inositol turnover in platelet activation; calcium mobilization and protein phosphorylation. Cell Calcium (3): 323-335, 1982.
118. Otani S, Matsui I, Kuramoto A, Morisawa S: Induction of ornithine decarboxylase in guinea-pig lymphocytes and its relation to phospholipid metabolism. Biochim Biophys Acta (800): 96-101, 1984.
119. O'Flaherty JT, Schmitt JD, McCall CE, Wykle RL: Diacylglycerols enhance human neutrophil degranulation responses: Relevancy to a multiple mediator hypothesis of cell function. Biochem Biophys Res Commun (123): 64-70, 1984.
120. Kajikawa N, Kaibuchi K, Matsubara, Kikkawa U, Takai Y, Nishizuka Y: A possible role of protein kinase C in signal-induced lysosomal enzyme release. Biochem Biophys Res Commun (116): 743-750, 1983.
121. Fujita I, Irita K, Takeshige K, Minakami S: Diacylglycerol, 1-oleoyl-2-acetyl-glycerol, stimulates superoxide-generation from human neutrophils. Biochem Biophys Res Commun (120): 318-324, 1984.
122. McNamara MJC, Schmitt JD, Wykle RL, Daniel LW: 1-O-Hexadecyl-2-acetyl-sn-glycerol stimulates differentiation of HL-60 human promyelocytic leukemia cells to macrophage-like cells. Biochem Biophys Res Commun (122): 824-830, 1984.
123. Feuerstein N, Cooper HL: Rapid phosphorylation-dephosphorylation of specific proteins induced by phorbol ester in HL-60 cells. J Biol Chem (259): 2782-2788, 1984.
124. Nishino H, Iwashima A, Nakadate T, Kato R, Fujiki H, Sugimura T: Potent antitumor promoting activity of N-(6-aminohexyl)-5-chloro-1-naphthalenesulfonamide, a calmodulin antagonist, in mouse skin tumor formation induced by 7,12-dimethylbenz[a]anthracene plus teleocidin. Carcinogenesis (5): 283-285, 1984.

125. Rasmussen H, Forder J, Kojima I, Scriabine A: TPA-induced contraction of isolated rabbit vascular smooth muscle. Biochem Biophys Res Commun (122): 776-784, 1984.
126. Kraft AS, Anderson WB, Cooper HL, Sando JJ: Decrease in cytosolic calcium/phospholipid-dependent protein kinase activity following phorbol ester treatment of EL4 thymoma cells. J Biol Chem (257): 13193-13196.
127. Kikkawa U, Takai Y, Minakuchi R, Inohara S, Nishizuka Y: Calcium-activated, phospholipid-dependent protein kinase from rat brain. J Biol Chem (257): 13341-13348, 1982.
128. Ogawa Y, Takai Y, Kawahara Y, Kimura S, Nishizuka Y: A new possible regulatory system for protein phosphorylation in human peripheral lymphocytes. J Immunol (127): 1369-1374, 1981.
129. Ku Y, Kishimoto A, Takai Y, Ogawa Y, Kimura S, Nishizuka Y: A new possible regulatory system for protein phosphorylation in human peripheral lymphocytes. J Immunol (127): 1375-1379, 1981.
130. Katakami Y, Kaibuchi K, Sawamura M, Takai Y, Nishizuka Y: Synergistic action of protein kinase C and calcium for histamine release from rat peritoneal mast cells. Biochem Biophys Res Commun (121): 573-578, 1984.
131. Cope FO, Staller JM, Mahsem RA, Boutwell RK: Retinoid-binding proteins are phosphorylated in vitro by soluble Ca^{2+}- and phosphatidylserine-dependent protein kinase from mouse brain. Biochem Biophys Res Commun (120): 593-601, 1984.
132. Iwasa Y, Hosey MM: Phosphorylation of cardiac sarcolemma proteins by the calcium-activated phospholipid-dependent protein kinase. J Biol Chem (259): 534-540, 1984.
133. Movsesian MA, Nishikawa M, Adelstein RS: Phosphorylation of phospholamban by calcium-activated, phospholipid-dependent protein kinase. J Biol Chem (259): 8029-8032, 1984.
134. Ahmad Z, Lee FT, DePaoli-Roach A, Roach PJ: Phosphorylation of glycogen synthase by the Ca^{2+}- and phospholipid-activated protein kinase (protein kinase C). J Biol Chem (259): 8743-8747, 1984.
135. Werth DK, Niedel JE, Pastan I: Vinculin, a cytoskeletal substrate of protein kinase C. J Biol Chem (258): 11423-11426, 1983.
136. Werth DK, Pastan I: Vinculin phosphorylation in response to calcium and phorbol esters in intact cells. J Biol Chem (259): 5264-5270, 1984.
137. Kawamoto S, Hidaka H: Ca^{2+}-activated, phospholipid-dependent protein kinase catalyzes the phosphorylation of actin-binding proteins. Biochem Biophys Res Commun (118): 736-742, 1984.
138. Schatzman RC, Grifo JA, Merrick WC, Kuo JF: Phospholipid-sensitive Ca^{2+}-dependent protein kinase phosphorylates the β subunit of eukaryotic initiation factor 2 (eIF-2). FEBS Lett (159): 167-170, 1983.
139. Hirata F, Matsuda K, Notsu Y, Hattori T, del Carmine R: Phosphorylation at a tyrosine residue of lipomodulin in mitrogen-stimulated murine thymocytes. Proc Natl Acad Sci USA (81): 4717-4721, 1984.
140. Imoka T, Lynham JA, Haslam RJ: Purification and characterization of the 47,000-dalton protein phosphorylated during degranulation of human platelets. J Biol Chem (258): 11404-11414, 1983.
141. Endo T, Naka M, Hidaka H: Ca^{2+}-phospholipid dependent phosphorylation of smooth muscle myosin. Biochem Biophys Res Commun (105): 942-948, 1982.
142. Nishikawa M, Hidaka H, Adelstein RS: Phosphorylation of smooth muscle heavy meromyosin by calcium-activated, phospholipid-dependent protein kinase. J Biol Chem (258): 14069-14072, 1983.
143. Naka M, Nishikawa M, Adelstein RS, Hidaka H: Phorbol ester-induced activation of human platelets is associated with protein kinase C phosphorylation of myosin light chains. Nature (306): 490-492, 1983.

144. Nishikawa M, Sellers JR, Adelstein RS, Hidaka H: Protein kinase C modulates in vitro phosphorylation of the smooth muscle heavy meromyosin by myosin light chain kinase. J Biol Chem (259): 8808-8814, 1984.
145. Couturier AN, Bazgar S, Castagna M: Further characterization of tumor-promoter-mediated activation of protein kinase C. Biochem Biophys Res Commun (121): 448-455, 1984.
146. Miyake R, Tanaka Y, Tsuda T, Kaibuchi K, Kikkawa U, Nishizuka Y: Activation of protein kinase C by non-phorbol tumor promoter, mezerein. Biochem Biophys Res Commun (121): 649-656, 1984.
147. Michell RH: Inositol phospholipids and cell surface receptor function. Biochim Biophys Acta (415): 81-147, 1975.
148. Kaibuchi K, Takai Y, Ogawa Y, Kimura S, Nishizuka Y: Inhibitory action of adenosine 3',5'-monophosphate on phosphatidyl inositol turnover: Difference in tissue response. Biochem Biophys Res Commun (104): 105-112, 1982.
149. Engelhard VH, Storm DR, Glaser M: Negative cooperativity exhibited by prostaglandin E_1 stimulated adenylate cyclase in LM cells. Biochemistry (17): 5304-5308, 1978.
150. Rao CV, Harker CW: Prostaglandin E and $F_{2\alpha}$ receptors in bovine corpus luteum plasma membranes are two different macromolecular entities. Biochem Biophys Res Commun (85): 1054-1060, 1978.
151. Fischer SM: The role of prostaglandins in tumor production. In: Slaga TJ (ed) Mechanisms of tumor promotion, Volume 2: Tumor promotion and skin carcinogenesis. CRC Press, Boca Raton, 1984, pp 113-126.
152. Kuehl FA, Egan RW: Prostaglandins, arachidonic acid, and inflammation. Science (210): 978-984, 1980.
153. Verma AK, Ashendel CL, Boutwell RK: Inhibition by prostaglandin synthesis inhibitors of the induction of epidermal ornithine decarboxylase activity, the accumulation of prostaglandins, and tumor promotion caused by 12-O-tetradecanoylphorbol-13-acetate. Cancer Res (40): 308-315, 1980.
154. Furstenberger G, Delescluse C, Fischer SM, Richter H, Marks F: Early induction of the arachidonic acid cascade and stimulation of DNA synthesis by TPA in murine and guinea pig epidermal cells in culture. In: Hecker E, Fusenig NE, Kunz W, Marks F, Thielmann HW (eds) Carcinogenesis a comprehensive survey, Volume 7: Cocarcinogenesis and biological effects of tumor promoters. Raven Press, New York, 1982, pp 325-329.
155. Slaga TJ: Multistage skin tumor promotion and specificity of inhibition. In: Slaga TJ (ed) Mechanisms of tumor promotion, Volume 2: Tumor promotion and skin carcinogenesis. CRC Press, Boca Raton, 1984, pp 189-196.
156. Levine L: Effects of tumor promoters on arachidonic acid metabolism by cells in culture. In: Hecker E, Fusenig NE, Kunz W, Marks F, Thielmann HW (eds) Carcinogenesis a comprehensive survey, Volume 7: Cocarcinogenesis and biological effects of tumor promoters. Raven Press, New York, 1982, 477-494.
157. Ashendel CL, Boutwell RK: Prostaglandin E and F levels in mouse epidermis are increased by tumor-promoting phorbol esters. Biochem Biophys Res Commun (90): 623-627, 1979.
158. Irvine RF: How is the level of free arachidonic acid controlled in mammalian cells? Biochem J (204): 3-16, 1982.
159. Klein-Szanto AJP: Morphological evaluation of tumor promoter effects on mammalian skin. In: Slaga TJ (eds) Mechanisms of tumor promotion, Volume 2: Tumor promotion and skin carcinogenesis. CRC Press, Boca Raton, 1984, pp 41- 72.
160. Klein-Szanto AJP, Major SK, Slaga TJ: Quantitative evaluation of dark keratinocytes induced by several promoting and hyperplasiogenic agents: Their use as an early morphological indicator of tumor-promoting action.

In: Hecker E, Fusenig NE, Kunz W, Marks F, Thielmann HW (eds) Carcinogenesis a comprehensive survey, Volume 7: Cocarcinogenesis and biological effects of tumor promoters. Raven Press, New York, 1982, pp 305-310.
161. Hirata F: The regulation of lipomodulin, a phospholipase inhibitory protein, in rabbit neutrophils by phosphorylation. J Biol Chem (256): 7730-7731, 1981.

Chapter 6

ACTIVE OXYGEN AND PROMOTION
Peter A. Cerutti
Swiss Institute for Experimental Cancer Research
Lausanne, Switzerland

Page

I. Introduction. 132

II. Induction of a cellular proxidant state 135
   A. Hyperbaric oxygen tension 136
   B. Radiation. 136
   C. Xenobiotic metabolism and Fenton-type reactions. 137
   D. Modulation of the cytochrome electron transport chain. . .138
   E. Peroxisome proliferators 139
   F. Inhibitors of the antioxidant defense systems. 139
   G. Membrane-active agents 140

III. Macromolecular damage resulting from cellular prooxidant
    states. 142
   A. Oxidative damage to membrane lipids. 142
   B. Oxidative damage to proteins 143
   C. Oxidative chromosomal damage 145

IV. Formation of chromosomal damage by the tumor promoter
    phorbol-12-myristate-13-acetate and the complete carcinogen
    aflatoxin B_1. 147
   A. PMA induces membrane-mediated chromosomal damage by
      indirect action. .147
   B. Membrane-mediated chromosomal damage by AFB_1 150

V. Concluding remarks. 152

References. 155

S.M. Fischer and T.J. Slaga (eds.), ARACHIDONIC ACID METABOLISM AND TUMOR PROMOTION. Copyright © 1985. Martinus Nijhoff Publishing, Boston. All rights reserved.

6. ACTIVE OXYGEN AND PROMOTION
PETER A. CERUTTI

I. INTRODUCTION

Evidence for the participation of active oxygen and organic hydroperoxides in carcinogenesis has accumulated over the last two decades. The major forms of active oxygen are superoxide (O_2^-) and its conjugate acid the hydroperoxyradical (HO_2^-), singlet oxygen ($^1O_2$), the hydroxyl radical ($^{\bullet}OH$) and hydrogenperoxide (H_2O_2). On the basis of epidemiological considerations, Totter (1) has proposed that oxygen-induced cellular damage rather than exposure to industrial pollutants may play a major role in human cancer. The hereditary chromosomal breakage disorders, Ataxia telangiectasia (AT), Fanconi's Anemia (FA) and Bloom's Syndrome (BS), are characterized by increased cancer incidence (9). There is evidence for abnormalities in the metabolism of oxygen for all three diseases (10). Cells from patients with these diseases possess increased spontaneous frequencies of chromosomal aberrations and, in the case of BS, also increased frequencies of sister chromatid exchanges (11) (SCE) and spontaneous mutations (12,13). The chromosomal abnormalities in BS fibroblasts can be decreased upon treatment with protease inhibitors (14). This is of particular interest because certain protease inhibitors suppress the induction of an oxidative burst by the tumor promotor phorbol-12-myristate-13-acetate (PMA) in polymorphonuclear leukocytes (15) and possess anticarcinogenic activity (16,17). Cultured cells from patients with these diseases are hypersensitive to agents which are known to induce a cellular prooxidant state, i.e. an increased flux of active oxygen. AT is sensitive to X-rays (18), bleomycin (19) and neocarcinostatin (20); FA to mitomycin C (21,22) and psoralens (23,24); BS to near-ultraviolet (25). Near-ultraviolet also induces excessive DNA strand breakage in BS-fibroblasts (26). Hyperbaric oxygen causes excessive chromosomal breakage in FA lymphocytes (27). Patients with AT and BS release clastogenic factors (CF) into their serum (28,29) and BS-(30) and AT-fibroblasts (28) produce CF in culture. CF are low molecular weight components of yet unidentified structure which break chromosomes in cultures of lymphocytes from healthy donors. The observation that Cu-Zn superoxide dismutase (SOD) inhibited the action of CF from BS indicates that O_2^- radicals are formed as intermediates in the clastogenic process. On the basis of these observations we have speculated that a chronic prooxidant state in these chromosomal breakage disorders may exert a promotional effect (31).

Experiments with whole animals and cultured animal cells support the notion that active oxygen plays a role in carcinogenesis. An elevation in oxygen tension increased the incidence of lung tumors in mice that had been fed dibenzanthracene (2) and increased the level of "spontaneous" transformation of cultured mouse embryo cells, particularly when they were exposed to fluorescent light (3). In our own work we have recently demonstrated that extracellularly generated superoxide (O_2^-) possesses strong promotional activity in mouse embryo fibroblasts 10T1/2 which had been initiated with γ-rays or benzo(a)pyrene-diolepoxide I (4). The carcinogenicity of organic peroxides on mouse skin (5) and the promotional activity of peroxyacetic acid on 7,12-dimethylbenz(a)anthracene (DMBA) initiated mouse skin (6) have been documented. Recently the promotional activity of an additional series of organic peroxides has been demonstrated (7,8).

Strong support for the intermediacy of active oxygen and peroxidatic reactions in carcinogenesis also derives from the anticarcinogenic activity of many antioxidants and radical scavengers (for reviews see 32,33,34). Shamberger and collaborators related the low mortality from gastrointestinal cancer in certain cities to high Se intake (35) (note : Se is a cofactor for glutathione (GSH)-peroxidase, the only known enzyme capable of destroying lipid and nucleic acid hydroperoxides (36)). The same authors demonstrated experimentally that Na$_2$SeO$_3$ and d,1-α- tocopherol (Vit E) reduced papilloma formation when administered before or concomittantly with croton oil to DMBA-treated mouse skin (37). Indeed, anticarcinogenic activity of a diet containing Se was observed as early as 1949 by Clayton and Baumann for animals which had been administered aminoazobenzene (38). The antioxidants VitE and VitC, inhibited tumor formation when they were administered to mouse skin shortly before the initiator (DMBA) in 2-step carcinogenesis experiments using PMA as the promotor (39). VitC and E also affected the expression of the malignant phenotype of cultured cells. Vit C reverted methylcholanthrene transformed foci of mouse embryo fibroblasts 10T1/2 at early stages to non-transformed morphology (40) and a Vit E derivative, D-α-tocopherol acid succinate, induced morphological changes and growth inhibition in mouse melanoma cells B16 and mouse L-cells (41). Extensive work on the anticarcinogenic action of the antioxidants butylated hydroxytoluene (BHT) and butylated hydroxyanisole (BHA) has been carried out by Wattenberg and collaborators (see in 32). In these experiments BHA and BHT may have inhibited the formation of covalent carcinogen DNA adducts by reducing the metabolic activation of compounds to ultimate carcinogens, e.g. by inhibiting cooxygenation reactions, and by increasing the elimination of the carcinogen(32,39). At the same time they may have acted as antipromotors. Slaga and collaborators found that they antagonized the promotional activity of PMA (39) and benzoylperoxide (8) in initiated mouse skin. Similarly, a biomimetic SOD (Cu^{2+}3,5-diisopropylsalicylic acid) suppressed the promotional action of PMA on mouse skin (42). SOD itself, but not catalase, partially inhibited the X-ray- and bleomycin-induced transformation of cultured hamster embryo cells and the promotional action of PMA (43). However, SOD had no effect on transformation in analogous experiments with mouse embryo fibroblasts 10T1/2 (44). This discrepancy may reflect inherent differences in the antioxidant defence of different cell strains and species. Uric acid is a strong antioxidant which is contained in human blood and saliva. It may play a role as a natural anticarcinogen (45). As discussed below there are several classes of exogenous agents which induce a cellular prooxidant state. Many of these are inflammatory agents, tumor promotors or complete carcinogens. Some cause aging related cellular degeneration. These cytopathological changes may be the consequence of ubiquitous oxidative damage to membrane lipids, to structural and catalytic proteins and to nucleic acids. The emphasis in this article is on the role of prooxidant states in tumor promotion.

II. INDUCTION OF A CELLULAR PROOXIDANT STATE

During evolution the change from anaerobic to aerobic metabolism brought about the advantage of increased amounts of energy from food but at the same time the disadvantage of oxygen toxicity,-mutagenicity and -carcinogenicity. These detrimental biological effects are due to the formation of small amounts of active oxygen species rather than dioxygen, O_2, itself (46,47). Some active oxygen is produced during the course of normal cellular metabolism. Large amounts can be produced as a result of the exposure of cells to exogenous physical and chemical agents and consequently oxygen homeostasis can be disturbed (48). However, cells are protected by an elaborate defence system consisting of small molecules that can react with active oxygen or secondary organic radicals formed from cellular target molecules. Non-protein sulfhydryls, alcohols and amines donate hydrogen, can scavenge ·OH and break radical chain reactions. Crucial physiological representatives of this group of compounds are cysteine, reduced glutathione (GSH) and Vitamins E and C. Polyunsaturated aliphatic molecules such as β-carotene scavenge $1_{O_2}*$ which adds to double bonds. Certain phenolic compounds react with organic hydroperoxyradicals, most importantly lipid-hydroperoxides, to yield harmless derivatives. Vitamin E breaks lipid-peroxidation chain reactions by this mechanism. In addition to these protector molecules of low molecular weight there are three types of enzymes which destroy active oxygen species : SOD dismutates O_2^- to $H_2O_2 + O_2$, catalases and peroxidases destroy H_2O_2 under formation of $H_2O + O_2$. Glutathione-peroxidase is also capable of destroying organic hydroperoxides formed from lipids and nucleic acids (36,48). A cellular prooxidant state, i.e. an excessive flux of active oxygen can establish itself when these antioxidant defence systems are deficient or overwhelmed by overproduction of active oxygen. The quality, quantity and intra- and extracellular distribution of active oxygen are not expected to be identical for prooxidant states produced via different mechanisms and by different types of cells.

Prooxidant states can be produced by the following major mechanisms:

A. Hyperbaric oxygen tension: At a concentration exceeding approximately 40% oxygen (i.e. dioxygen, O_2) becomes toxic to most tissues (49). After a remarkably long period, cell division is delayed and macromolecular synthesis is inhibited (50). High oxygen tension inhibits initiation- and elongation-steps in DNA synthesis probably because of damage to DNA and the replication machinery (51). It also induces mutations and chromosomal aberrations (3,52,53,54). Active oxygen species most likely play a role in mediating these effects. The ample evidence which exists for the cytotoxicity, mutagenicity, clastogenicity (55, 56, 57, 58, 59) and carcinogenicity (2,3,4) of these reactive species supports such a notion.

B. Radiation: Ionizing radiation and near-ultraviolet light (near-UV) can produce cellular prooxidant states, by forming O_2^-, $^1O_2^*$ and ·OH. Consequently, ionizing radiation induces DNA damage mostly by indirect action of these active oxygen species (60, 61,62, 63). It is mutagenic, clastogenic and carcinogenic. For near-UV, the contribution of indirect action to the formation of damage increases with increasing wavelengths (64). Near-UV is mutagenic (65, 66, 67), clastogenic (3) and carcinogenic (3, 67, 68) for cultured cells and there is convincing epidemiological data (69) that the near-UV component of solar radiation is largely responsible for non-melanoma skin cancer in the human. Indeed, the combination of direct and indirect mechanisms in the formation of DNA damage may render UV-B light (wavelenghts 290-320 nm) the potent <u>complete</u> carcinogen that it is. Photosensitization by xenobiotics results in $^1O_2^*$ formation, (e.g. by polycyclic aromatic hydrocarbons (PAH) and other airpollutants (70), iatrogenic psoralens (24) etc.) and may play a role in skin carcinogenesis. There are volumes of radiation-chemical and radiation-biological literature which describe the effects of radiation on whole organisms, cells and molecules (71,72).

C. Xenobiotic metabolism and Fenton-type reactions: There are numerous xenobiotics which produce active oxygen spontaneously or via metabolic activation (73, 74). Many of these compounds are promotors or complete carcinogens. (a) Certain <u>quinoid molecules</u> can participate in redox cycles and transfer an electron to O_2 under formation of O_2^- when they are oxidized from semiquinones to quinones (75). It should be kept in mind that not only active oxygen but also the xenobiotic radicals themselves can react with the cellular target molecules. Major representatives of this group of compounds are the carcinostatic (and carcinogenic) benzanthraquinones (e.g. adriamycin and daunorubicin), the N-heterocyclic quinones (e.g. streptonigrin and mitomycin C) and certain PAH. Quinone metabolism is associated with pyridine nucleotide oxidation (76). Because NADPH oxidation is responsible for the release of mitochondrial Ca^{2+} (77), quinoid metabolism can disturb the homeostasis of cellular Ca^{2+}. This can have important consequences for cytoskeletal structure and function (78). Support for the intermediacy of active oxygen in the mechanism of action of these chemicals derives from the fact that antioxidants inhibit benzanthraquinone induced lipidperoxidation (79) and daunomycin-(80) and PAH-induced carcinogenesis (81). (b) Many <u>nitroaromatic compounds</u> are reduced by nitroreductases to hydroxylamines via the intermediate formation of free radicals. These free radicals can transfer an electron to O_2 as they revert to starting material. The result is the catalytic formation of O_2^- (73). An important carcinogenic representative of this group of compounds is 4-nitroquinoline-N-oxide (4-NQO) (82). The observation that GSH prevents 4-NQO-induced cytotoxicity implicates radical reactions (83, 84). In addition to DNA damage induced by indirect action 4-NQO also forms covalent adducts to DNA (85). (c) <u>Fenton-type reagents</u> produce ˙OH-radicals in reactions involving Fe- or Cu- ions. H_2O_2, ascorbate and organic hydroperoxides (in a broad sense) belong to this class of compounds. In these reactions, organic hydroperoxides produce alkoxy-radicals which can cause branching in lipid oxidation chains. The anticancer drug bleomycin (86) can be considered as a Fenton-reagent which is targeted to DNA. Fenton-reagents induce DNA strand breakage with high efficiency (87,88). (d) <u>Others</u>: azo-anion, hydrazyl- and bipyridylium-radicals can transfer an electron to O_2 and potentially induce a cellular prooxidant state (73). Typical representatives are sulfonazo III (89), phenylhydrazine (90) and paraquat (91). They can induce lipid-peroxidation and hemolysis (92, 93). The trichloromethyl radical derived from CCl_4 is an efficient initiator of lipid-peroxidation chain reactions (94) and a potent hepatotoxin and hepato-carcinogen (95). CCl_4 acted as hepato-promotor when it was administered to animals which had received an initiating dose of neutron radiation (96). Similarly, the environmental toxin 2, 3, 7, 8-tetrachlorodibenzo(p)dioxin (TCDD) is a carcinogen for rats and mice (95) and a promotor for rat liver (97) and skin of hairless mice (98). It produces lipidperoxidation in rat liver (99).

D. <u>Modulation of the cytochrome electron transport chain:</u>   The mitochondrial and microsomal cytochrome electron-transport chains release active oxygen under physiological conditions (100,101,102). Lipidhydroperoxides are formed and subsequently degraded by cytochrome P450 peroxidase (103). Active oxygen that is released in the peroxidase reaction can inactivate cytochrome P450 (104, 105). The amount of active oxygen that is produced depends on the respiratory state of the organelle, the presence or absence of inhibitors, uncouplers, inducers, and the type of substrate which is being metabolized (101, 103). For example the inhibitors antimycin and rotenone enhance active oxygen flux (100) as do certain substrates and inducers of microsomal cytochrome P450 metabolism (102). It is interesting to note that rotenone is a hepatocarcinogen (106) and that the P450 inducer, phenobarbital, is a potent hepatic tumor promoter (107).

E. <u>Peroxisome proliferators:</u> Peroxisomes are ubiquitous cytoplasmic organelles which contain catalase, H_2O_2 generating oxidases, as well as enzymes involved in the ß-oxidation of long chain fatty acids (108,109,110). There are a number of structurally unrelated hypolipidaemic drugs (e.g. clofibrate, nafenopin) which induce the proliferation of these organelles as well as hepatomegaly in rodents. They are hepatocarcinogenic or hepatococarcinogenic (111). The hepatic promotor and initiator, di(2-ethylhexyl)phtalate (DEHP) may owe its activity to two mechanisms. DEHP is a peroxisomal stimulator and its metabolite monoethylhexylphtalate is an inhibitor of the electron transport chain (112). Thus as a peroxisomal proliferator and inhibitor of mitochondrial electron transport DEHP may induce genetic damage by indirect action due to an increase in the intracellular concentration of active oxygen. There is no evidence that these drugs form adducts with DNA and they are non-mutagenic in the S9 supplemented Salmonella assay (113).

F. <u>Inhibitors of the antioxidant defense systems</u>: Agents that reduce the detoxication capacity of the major enzymes which protect cells from active oxygen and that diminish the concentration of the low molecular weight scavenging molecules can result in a cellular pro-oxidant state. Examples include inhibitors of catalase such as azide, hydroxylamine, aminotriazole (114,115) and thiocarbamatic acid derivatives (116).The latter group of compounds are especially known to inhibit CuZn SOD and apparently GSH-peroxidase(116). These thiocarbamates act as metal chelators but also react with protein- and non-protein-SH-groups. Because they also decrease the concentration of GSH they drastically diminish the cellular antioxidant capacity (117, 118). Experiments with cultured cells and whole animals support the view that the biological effects of dithiocarbamates are due to the formation of a prooxidant state (119). The mutagenicity of these compounds in Salmonella typhimurium is increased by oxygen and the addition of microsomes plus a NADPH-generating system (120). The skin tumor promotor PMA decreases the SOD and CAT activities in mouse epidermis and may by this mechanism induce or exacerbate a prooxidant state (121). A majority of tumor cells appear to possess decreased levels of Mn-SOD. It has been speculated that the resulting weakened antioxidant defense in tumor cells may play a role in the carcinogenic process (122).

GSH represents the most important cellular antioxidant. It is actively transported into the extracellular milieu by membrane-bound γ-glutamyl transpeptidase. Therefore, it also protects cellular membranes from radicals that are generated in the extracellular space (123,124). Hydroperoxides, diamide, diethylmaleate and 1-chloro-2,4-dinitrobenzene depress GSH levels (125). However, none of these compounds are specific for GSH. Prothionine sulfoximine and buthionine sulfoximine decrease intracellular GSH levels selectively by inhibiting γ-glutamyl cysteine synthetase (123). Treatment of mouse tumor cells and cultured human lymphoid cells with these drugs increased cellular sensitivities to cytolysis by peroxide (126) and to lethality by radiation (127), respectively.

G. Membrane-active agents: There are agents of xenobiotic- and endogenous origin and of different molecular weights and structure which share the property of membrane activity. They can affect plasma-, nuclear-, mitochondrial-, Golgi- or endoplasmic reticulum-membranes and interact with receptors or perturb their conformational integrity in a less specific manner. Examples of membrane-active agents of particular importance to carcinogenesis are : peptide hormones, growth factors, lectins, the tumor promotors PMA (128, 129, 130, 131, 132), teleocidin (133) and mezerein, complete carcinogens such as aflatoxin B_1 and benzo(a)pyrene (129, 134, 135), certain bacteria and viruses (136), particulates (137, 138) such as asbetos and silica, as well as components of the immune system (139, 140).

Depending on the membrane-active agent and the cell type pro-oxidant states can be induced by different mechanisms. Perturbation of membrane-conformation by chaotropic agents may render membrane lipids susceptible to autoxidation (141). Other agents stimulate membrane phospholipases (C or A_2) resulting in the formation of diacylglyceride, free arachidonic acid (AA) and increased amounts of AA-metabolites (31, 142). Biosynthesis of prostaglandins (PG) and hydroxy-AA proceeds via the intermediate formation of the hydroperoxy-derivatives PG G_2, 15-hydroperoxy-PGE_2 and 5, 12, or 15-hydroperoxy-AA. These unstable intermediates react to the corresponding hydroxyl-derivatives spontaneously or in peroxidase catalyzed reactions and release active oxygen (143, 144, 145). In specialized phagocytic leukocytes membrane active agents can elicit an oxidative burst by the activation of a NADPH dependent oxidase. Most of the active oxygen produced in this fashion is released into the extracellular space and plays a crucial role in the bactericidal action of these cells (146). Natural killer cells may react in a similar fashion (147). The activation of phospholipase A_2 and NADPH-dependent oxidase may be the consequence of conformational changes in the membrane or of a kinase cascade initiated by protein kinase C.

The membrane-active tumor promotor PMA directly stimulates protein kinase C which acts as PMA receptor (148, 149, 150). It stimulates the release of free AA and the formation of AA-metabolites in many different types of cells including human leukocytes. PMA induces the formation of clastogenic factors (CFs) in human monocytes, polymorphonuclear leukocytes and "regular" lymphocyte preparations which are contaminated with monocytes and platelets (151). (CFs are low molecular weight components which break chromosomes in fresh blood- or lymphocyte test cultures). We speculate that these CFs consist of hydroperoxy-AA derivatives and their aldehydic degradation products plus free AA and that autoxidation chain reactions are involved in the clastogenic mechanism (31). Depending on the cell type the hydroperoxy-AA derivatives may be 15-hydroperoxy-PG E_2, PG G_2, 5, 12 or 15-hydroperoxy-AA. For human monocytes it probably consists of 15-hydroperoxy-PG E_2 (see below). Clastogenic lipid peroxides could also be produced without the stimulation of the AA-cascade as a consequence of the preferential degradation of oxidized membrane-phospholipids by phospholipase A_2 (36). Membrane-active agents can induce chromosomal damage in the target-cell itself but also produce damage in surrounding and remote tissues via the intermediacy of CFs (31, 142).

III. MACROMOLECULAR DAMAGE RESULTING FROM CELLULAR PROOXIDANT STATES

A. Oxidative damage to membrane-lipids

Membrane phospholipids are particularly susceptible to oxidative attack. In the presence of oxygen non-enzymatic reactions appear to be catalyzed by a transition metal ion and a reducing agent. However, some oxidation also occurs in the absence of metal ions. The polyunsaturated fatty acid substituents are oxidized in radical chain reactions to peroxides. Lipid-hydroperoxides and -endoperoxides are of limited stability. They can decay to alkoxy radicals which reinitiate radical chains. Alternatively they can fragment to aldehydic products such as malonedialdehyde and 4-hydroxy-alkenals, rearrange to hydroxy- and keto-acids or react to complex polymerized mixtures (152, 153, 154). Membrane conformation and function may be impaired and the activity of membrane bound enzymes altered. Reactive radical derivatives and aldehydic degradation products of lipidperoxides may potentially attack other macromolecular targets in the same cell and conceivably also in remote tissues (155). They may induce chromosomal damage, i.e. act as clastogenic factors (CF) (31, 142). Lipidperoxides in membranes may attack neighboring proteins and cause their inactivation or changes in their function (156).

Glycolipids such as gangliosides are readily oxidized at their sialic acid residues which are converted to aldehydes. Free aldehyde formation on the cell surface may play a crucial role in blastic transformation of lymphocytes by various oxidizing agents (e.g. IO_4, galactose oxidase (157)). Oxidation of the cell surface trisialo-ganglioside GT_1 may be involved in the promotion of the neoplastic transformation of JB6 mouse epidermal cells by IO_4, benzoylperoxide and PMA (158). Chemical considerations would predict that different products are formed from glycolipids when reacted with IO_4, which attacks vicinal hydroxyl groups, organic hydroperoxides and active oxygen species.

Lipid oxidation can be prevented by eliminating the initiating active oxygen species in free radical chain reactions. This can be achieved by the enzymes SOD for $\overline{O_2}$ and catalase and certain peroxidases for H_2O_2 (159). GSH-peroxidase is unique because it can degrade lipid hydroperoxides (36) which otherwise could participate in chain-branching reactions. Active oxygen can also be scavenged by non-protein-sulfhydryls such as cysteine, cysteinyl-glycine and GSH. Non-protein-sulfhydryls and phenolic compounds such as VitE and BHA also break lipid peroxidation chains. They donate hydrogen to radical intermediates and react themselves to harmless secondary products (160, 161). VitE is mostly associated with membrane lipids, approximately one third is found in endoplasmatic reticulum and two thirds in mitochondria (162). The capacity of these antioxidants to diminish lipid oxidation may be important for their anticarcinogenic properties (32, 33).

B. <u>Oxidative damage to proteins</u>

The radiation chemistry of proteins in aerobic solution gives an indication of the type of oxidative damage induced by active oxygen (72). Methionine in newly systhesized proteins is particularly sensitive to oxidation (163). For example two thirds of the methionine residues were oxidized to the sulfoxide following stimulation of an oxidative burst in human neutrophils by PMA (164). As mentioned above lipidperoxides may mediate oxidative damage to other macromolecules. Inactivation of cytochrome P450's by this mechanism could alter the metabolism of carcinogens and modulate their activity (165). In the present context oxidative damage to enzymes which participate in antioxidant defence is of particular importance because their inactivation is expected to potentiate the deleterious effects of a prooxidant state. An interesting synergism exists for catalase and CuZn and Fe SOD which mutually protect each other. Catalase is (partially) protected by SOD from inactivation by O_2^- and SOD is protected by catalase from inactivation by H_2O_2. These reactions have been documented <u>in vitro</u> but could play a role <u>in vivo</u> (166). GSH-peroxidase probably represents the most important enzyme in antioxidant defense because it not only destroys H_2O_2 (Se enzyme at low H_2O_2 concentrations) but also hydroperoxides of lipids (36) and nucleic acids (167). The activity of GSH-peroxidase depends on the intracellular GSH concentrations. Depletion of the cellular reducing capacity as a consequence of a prooxidant state can have a disastrous effect not only because insufficient non-protein-SH is available for the prevention of initiation and propagation of radical chain reactions (e.g. lipid peroxidation chains), but also because of the diminution of the activity of GSH-peroxidase (36).

Maintenance of the structural integrity of DNA is extraordinarily important to the cell and a complex enzymatic system for the repair of damage to DNA has evolved. Oxidative damage to DNA repair enzymes would potentiate the deleterious effects of a prooxidant state on the genome. A case in point is the oxygen-dependent killing of E.Coli with 365 nm near-UV. Only the uvrA recA double mutant lacked the oxygen effect suggesting that damage to repair enzymes was responsible for the oxygen-dependent killing of the wildtype (168). It is conceivable that aldehydes produced in fragmentation reactions from oxidized lipids inactivate repair enzymes. Inhibition of repair by formaldehyde has been observed (169).

C. Oxidative chromosomal damage

The recognition by E. & J. Miller (170) that the covalent binding to DNA of many carcinogens requires their activation to electrophilic, ultimate metabolites was of pivotal importance to carcinogenesis research. We refer to this mechanism of DNA damage induction as "direct action". Several years ago we proposed to distinguish a second basic mechanism which we named "indirect action". Agents operating by indirect action produce secondary DNA damaging agents in reactions with cellular molecules other than DNA (171). The secondary agents are mostly active oxygen species, lipid-hydroperoxides (172) and their radical- and aldehydic degradation products (31, 152, 153), N-chloroamines (173), oxidation products of aromatic amino acids (174) and purines (175) etc. Physical and chemical agents which have the capacity to induce a prooxidant state are expected to cause DNA damage by indirect action. Characteristic DNA-lesions include strand breaks, products of the 5,6-dihydroxy-dihydrothymine type, apurinic- and apyrimidinic-sites etc.. Because breaks are also produced in the course of repair, the demonstration of DNA strand breakage does not suffice to implicate indirect action of chemical agents without rigorously excluding the formation of covalent adducts. Certain adducts can be formed via radical intermediates (73) and cooxidative metabolism (176). Therefore, the suppression of mutations or aberrations by radical scavengers also does not prove the involvement of indirect action. The only unambiguous proof is the demonstration of oxidative base damage, e.g. products of the 5,6-dihydroxy-dihydrothymine-type or formation of ^3H-H_2O from thymine-methyl-^3H. This has been accomplished for γ-rays (62, 63), near-UV (177), ascorbate-Cu^{2+} and benzo(a)pyrene (88). Strong indications for indirect action have also been obtained for aflatoxin B_1 (AFB_1) (178), and PMA (151, 179, 180) (see below). The availability of monoclonal antibodies against 5,6-dihydroxy-dihydrothymine containing DNA (181) opens a new avenue to the study of indirect action in carcinogenesis.

Active oxygen produces DNA damage and as a consequence mutations and chromosomal aberrations. Agents inducing a prooxidant state are usually extremely toxic. The demonstration of the formation of mutations and chromosomal aberrations by such agents requires fine-tuning of experimental conditions. Good cell growth in culture media with low antioxidant capacity is a prerequisite. Experimental difficulties are responsible for slow progress in the study of the genotoxic effects of active oxygen and for controversies in the literature (182). O_2^- produces DNA strand breakage (183, 184), mutations in Salmonella (55) and chinese hamster ovary cells (56), chromosomal aberrations in human lymphocytes (57) and is a weak complete carcinogen but potent promotor for mouse embryo fibroblasts C3H/10T1/2; H_2O_2 produces DNA strand breakage (185) and inactivates and (weakly) mutates B. subtilis transforming DNA (186) and Salmonella (58); H_2O_2 also induces strand breakage (59), sister chromatid exchanges (187), cytotoxicity but no point mutations in V79 cells (59). These effects of H_2O_2 are probably due to the formation of $^{\bullet}OH$ as the ultimate damaging species in Fenton-type reactions. Dioxygen (O_2) especially under hyperbaric conditions, induces mutations in bacteria (53), chromosomal aberrations in plant-(188) and mammalian cells (3, 52, 54) and point mutations (52) in mammalian-cells. O_2 probably exerts these effects because of the formation of small amounts of active oxygen by cellular metabolism. In summary, active oxygen species possess the following characteristics regardless of the mechanism of their formation: they are weak point mutagens, weak inducers of SCE but potent inducers of chromosomal aberrations. Chromosomal aberrations consist mostly of gaps and chromatid breaks.

As already mentioned, in addition to active oxygen species themselves oxidation products of non-DNA cellular molecules can inflict chromosomal damage. They can induce damage not only in the cell of their origins but also in surrounding tissue, i.e. they can act as diffusible mutagens or clastogenic factors (CF). CFs have been detected in the serum of patients with the cancer prone diseases AT (28) and BS (29) as well as in the serum of patients with certain rheumatic and autoimmune diseases (189). A CF is induced in human leukocytes in response to PMA treatment (151). As mentioned above, candidates for CFs are lipidperoxides (172) and their radical- and aldehydic degradation products (31, 152, 153) oxidation products of aromatic aminoacids (174) and purines (175), N-chloroamines (173) etc..

The formation of oxidative chromosomal damage can be suppressed by antioxidants. Non-protein-SH compounds, catalase and SOD reduce the efficiency of aberration induction by X-rays (190, 191) and O_2^- (57). In contrast to other macromolecules the integrity of DNA is further safeguarded by elaborate enzymatic repair systems. The following major repair pathways exist for DNA damage induced by indirect action: (1) repair of strand breaks (endpreparation (?), gap-filling, ligation) (2) destruction of hydroperoxides of DNA bases (GSH-peroxidase) (167) (3) base excision repair (glycosylases and AP-endonucleases). Some of these repair systems are apparently inducible (192). The existence of nucleotide excision repair, cross-link repair and post-replication repair of damage induced by indirect action has not been clearly demonstrated. Excellent recent reviews on DNA repair are available (193, 194).

IV. FORMATION OF CHROMOSOMAL DAMAGE BY THE TUMOR PROMOTOR PHORBOL-12-MYRISTATE-13-ACETATE AND THE COMPLETE CARCINOGEN AFLATOXIN B1

A. PMA induces membrane-mediated chromosomal damage by indirect action.

Certain membrane-active chemicals are capable of inducing chromosomal damage via indirect action. They damage chromosomes because they induce a prooxidant state. We have focussed in our own studies on the formation of chromosomal damage in human lymphocytes by the tumor promotor PMA. There is no evidence that it binds covalently to DNA. Ne-

vertheless, it induces chromosomal aberrations in human lymphocytes
(see below and 151, 179, 180) and mouse epidermal cells (195), aneuploidy in yeast and DNA SS-breaks (196) in human leukocytes enriched
in polymorphonuclear cells (PMN) (197). PMA also "facilitates" the amplification of dihydrofolate reductase genes in mouse 3T6 cells
(198). (Note: Similar amplification of this gene is obtained with the
bona fide DNA damaging agent ultraviolet light and with hydroxyurea
(199, 200) and may be a consequence of DNA replication on a damaged
template. Hydroxyurea may induce DNA damage by indirect action (201).
DNA damage that is induced by indirect action may affect the initiation of replication of large chromosomal domains. In analogy to ionizing radiation (202, 203), which represents the prototype of an indirect agent, PMA preferentially inhibits initiation of DNA replication (204).

PMA is a potent inducer of DNA single strand-breaks in pure PMNs
but a poor inducer of breaks in "regular" human lymphocyte preparations which contain monocytes and some platelets but very few PMNs
(205). This difference in the efficiency of strand breakage by PMA in
pure PMNs and regular lymphocytes may reflect different mechanisms.
Human PMNs but not lymphocytes respond with a massive oxidative burst
to PMA treatment. It is likely that the PMA induced prooxidant state
in PMN is responsible for strandbreakage in these cells. In regular
lymphocyte preparations the low level of strand breakage by PMA may
result from active oxygen released by monocytes, by a minor oxidative
burst from certain subclasses of lymphocytes (147) and the induction
of a prooxidant state via the stimulation of the AA-cascade (31, 142,
206, 207, 208).

PMA induces chromosomal aberrations in PHA stimulated human lymphocyte cultures (179). Mitogenic stimulation by PHA is a prerequisite
for the formation of aberrations by PMA. Efficient induction of aberrations is observed for platelet-free, relatively "pure" lymphocyte
cultures which are contaminated with maximally 1% PMN and 2% monocytes
(206). The intermediacy of active oxygen in the formation of aber-

rations is indicated by the anticlastogenic action of antioxidants (179, 180). The anticlastogenic activity of inhibitors of the arachidonic acid (AA) cascade suggests the involvement of AA-metabolism (180). However, it has to be kept in mind that many such inhibitors are also potent antioxidants.

PMA induces the release of a CF from regular human lymphocyte cultures (151). The CF consists of low molecular weight material which is released into the culture medium and has the capability of inducing chromosomal aberrations in fresh (non PMA-treated) blood or lymphocyte test cultures. However, "pure" lymphocytes containing no platelets and maximally 1% PMN and 2% monocytes, produced a very weak CF while potent CF preparations were obtained from PMA-treated pure monocytes and PMNs (206). Interleukin mediated intercellular interactions may augment CF production in lymphocyte preparations containing monocytes (209,210). We speculate that CFs induced by PMA in human leukocytes consist of H_2O_2, AA-peroxides and their aldehydic degradation products, unknown AA-metabolites (31), free AA, N-chloroamines (173) and possibly oxidation products of aromatic amino acids (174) and purines (175). Ethylacetate extracts of PMA treated human monocytes contained increased amounts of thromboxane B_2 (Tx B_2), 12-L-hydroxy-5,8,10-heptadecatrienoic acid (HHT), hydroxy-AA's and free AA (W. Kozumbo, D. Muhlematter, P. Amstad and P. Cerutti, unpublished). It is evident that PMA stimulates both the cyclooxygenase and lipoxygenase pathways of the AA-cascade. The biosynthetic formation of TxB_2 and HHT proceeds via the intermediacy of PGG_2 and is accompanied by the release of active oxygen and malondialdehyde. PGG_2, hydroperoxy-AA, H_2O_2 and OH-radicals produced in Fenton-type reactions may start autoxidation chain reactions in the target cell and cause chromosomal damage. Chromosomal damage in mouse epidermal cells may be induced by a similar mechanism because PMA-stimulated PGE_2 synthesis and AA-release (130, 211, 212). PGE_2 is formed from PGG_2 and 15-hydroperoxy-PGE_2 under release of active oxygen. The oxygen-dependent formation of two unidentified highly lipophilic AA-derivatives upon PMA treatment of bovine lymphocytes has recently been described (213). DNA strand breakage in mixed leukocytes was suppressed by catalase but not SOD (197). The latter results implicate an oxidative burst and H_2O_2 in the formation of DNA damage under these conditions. The situation is more complicated for the induction of chromosomal aberrations. SOD but not catalase was anticlastogenic for PMA-treated blood cultures (179) but the formation of CF from monocyte- or PMN-suplemented "pure" lymphocytes could be suppressed by both SOD and catalase (206).

CF may represent intra- and intercellular signals which translate events from the cell surface to the genome (31). This concept is supported by cocultivation experiments. Mixed human leukocytes and "regular" lymphocytes but not "pure" human lymphocytes produce mutations in Salmonella (214). PMA stimulated human leukocytes to induce SCE in monolayers of chinese hamster ovary cells (215) and DNA strand breakage in mouse erythroleukemia cells (216). Leukocytes from patients with chronic granulomatous disease were not mutagenic and did not induce SCE in CHO cells in response to PMA (215).

B. Membrane-mediated chromosomal damage by AFB_1

On the basis of our experiments with mouse embryo fibroblasts 10T1/2 (217, 218), epithelioid human lung cells A549 (219), fetal human hepatocytes (220) and human lymphocytes we conclude that the complete carcinogen AFB_1 induces chromosomal damage by three mechanisms: (1) formation of covalent DNA adducts via metabolic activation by the cytochrome P450 system (2) adduct formation via cooxygenation in conjunction with AA-metabolism (3) damage by indirect action because of its membrane active character. The contribution of cooxygenation to adduct formation was assessed in 10T1/2 cells by the use of the inhibitors of prostaglandin endoperoxide synthetase and lipoxygenase, indomethacin and 5,8,11,14-icosatetraynoic acid (ETYA). Indomethacin and ETYA inhibited AFB_1-binding to maximally 60%. For the interpretation of these results it should not be ignored that indomethacin is also a potent antioxidant (221). The antioxidant GSH was also inhibitory while CuZn SOD had no effect or slightly stimulated binding at high concentrations. The observation that the phospholipase A_2 inhibitor p-bromophenacylbromide diminished AFB_1-DNA binding supports the notion that AFB_1, because of its membrane-activity, enhances its own co-oxidative metabolism by stimulating the AA-cascade (218). The remaining 40% of AFB_1-DNA adduct formation presumably occurs by the cytochrome P450 system in 10T1/2 cells. The relative contribution of the two pathways is expected to vary for different cell types, e.g. it is likely that rat liver microsomes preferentially use the P450 system but ram seminal vesicle microsomes use prostaglandin synthetase.

Evidence for the formation of DNA damage by indirect action derives from experiments with phytohemagglutinin stimulated human lymphocytes (178). Only very few or no covalent AFB_1-DNA adducts were detectable in lymphocyte DNA after incubation for 20h with 0.3 µM AFB_1 while 0.1 µM AFB_1 sufficed to induce 19.4 ± 7.5% mitosis with aberrations. Inhibitors of AA-metabolism including the phospholipase A_2 inhibitor bromophenacylbromid and antioxidants were anticlastogenic. In particular, CuZn SOD was strongly anticlastogenic while it had no effect on the formation of covalent DNA adducts in 10T1/2 cells (218) (the radioactivity associated with DNA extracted from ^3H-AFB_1 treated lymphocytes was too low to study the effect of SOD on adduct formation). Additional support for the membrane-active character of AFB_1 derives from the observation that AFB_1 stimulated the release of free AA and hydroxy- and/or hydroperoxy-AA from "regular" lymphocytes. This is reminiscent of previous results with canine kidney cells. Benzo(a)pyrene stimulates the release of free AA and prostaglandin $F_2\alpha$ from these cells (222). Unambiguous proof for indirect action by AFB_1 requires the demonstration of characteristic lesions in DNA, e.g. of thymine damage of the 5,6-dihydroxy-dihydrothymine type.

V. CONCLUDING REMARKS

The evidence is convincing that agents capable of inducing a prooxidant state can be carcinogenic. It is no easy task to relate the effect of active oxygen to a specific step in malignant transformation, however. As discussed above, experimental results suggest a role in late steps, i.e. promotion and progression. Some of the examples mentioned above are (a) O_2^- promotes initiated mouse embryo fibroblasts 10T1/2 (4); IO_4^- promotes mouse epidermal cells JB6 (157); (b) H_2O_2, and various organic hydroperoxides possess promotional activity (7, 8, 158); (c) CCl_4 (96) and TCDD (97,98) which induce lipidperoxidation (94,99) possess promotional activity; (d) Phenobarbital (107) and DEHP (112) which increase the flux of active oxygen from cytochrome P450 are hepatopromotors; (e) The mouse skin promotors PMA and teleocidin induce a prooxidant state because of their membrane active properties (15, 31, 142, 179, 180); PMA also diminishes the antioxidant defense (121); (f) Many antioxidants are antipromotors and anticarcinogens, etc.. In the frame of the 2-stage promotion model of Slaga and collaborators (223) active oxygen probably qualifies as a first stage promotor.

Active oxygen, regardless of the mechanism of its formation, can induce macromolecular damage. As emphasized above, it damages membranes, proteins and nucleic acids and modifies their functional properties; it also decreases the cellular reducing power and disturbs Ca^{2+} homeostasis and consequently the structure and function of the cytoskeleton etc.. While any or all of these changes could have a promotional effect, damage to DNA is expected to have the farthest reaching consequences because it may result in temporary or permanent changes in gene expression. The dogma that (early) effects of tumor promotors are "fully" reversible argues against the notion that they cause structural genomic changes. This dogma, however, has recently been shaken. Two-stage promotion experiments on initiated mouse skin indicate that a single application of PMA induces a long lasting change. Promotion of such cells could be completed several weeks later by the second stage promotor 12-O-retinoylphorbol-13-acetate (224). PMA also induces an irreversible effect on cultured preneoplastic mouse epidermal cells JB6 which it promotes to the transformed phenotype (225, 226). PMA alone can induce malignant tumors on non-initiated mouse skin (227). A large degree of reversibility of the biological effects of agents inducing oxidative damage does not in fact argue against the involvement of chromosomal damage. DNA lesions induced by indirect action are repaired very rapidly and completely (see e.g. 88, 193, 194, 228, 229) and only the accumulation of a small fraction of irreparable lesions following repeated treatments may lead to permanent genetic changes. Evidence for the intermediacy of active oxygen in the induction of genes that are related to promotion comes from the observation that antioxidants suppressed the induction by PMA of ornithine decarboxylase in mouse epidermis (230) and mouse mammary tumor cells (231). Catalase also suppressed the induction by PMA of exogenous copies of mouse mammary tumor virus in the latter cells (J. Friedman and P. Cerutti, unpublished).

Genetic and genetic/epigenetic models can be considered for the modulation of gene expression by a cellular prooxidant state. (1) genetic model: DNA damage is induced by indirect action and in the processing of this damage by repair, replication and recombination permanent alterations occur to the DNA sequence, i.e. mutations, amplification of certain sequences, intra- and inter-chromosomal rearrangements of blocks of sequences. These events may result in the production of mutated proteins with altered activities and over- or under-production of proteins. Because DNA damaging agents which operate by indirect action are strong clastogens but weak mutagens, they may preferentially induce sequence rearrangements. Examples for all the above mechanisms of modulation of gene expression are available from oncogene research (232, 233). Alternatively, DNA strand breaks may modulate the expression of genetic domains by altering chromatin conformation without causing sequence changes. (2) A genetic/epigenetic model: a prooxidant state may modulate gene expression via poly (ADP)-ribosylation of chromosomal proteins. The cellular demand for detoxication of active oxygen and organic radicals is increased in a prooxidant state. As a result GSH levels decrease, e.g. as a consequence of the action of GSH-peroxidase, and the levels of the oxidized pyridine nucleotides, $NADP^+$ and NAD^+, increase. NAD^+ is the substrate for poly (ADP-ribose)synthetase. A prooxidant state induces DNA strand breaks, which stimulate poly (ADP-ribose)-synthetase (234). Therefore polymerase stimulation coupled with the increased availability of NAD^+ may result in the rapid poly ADP-ribosylation of chromosomal proteins and, consequently,the modulation of gene expression. Later, when poly (ADP-ribose)-polymerization is proceeding at a high rate NAD^+ levels drop. A distinct group of "prooxidant genes" may be affected that are analogous to the induction of heat shock genes in eukaryotes and SOS-functions in bacteria. The following genes and regulatory sequences may fall into the prooxidant category: ornithine-decarboxylase, plasminogen activator, poly(ADP-ribose)-polymerase, genes of cyclic nucleotide metabolism, genes of the antioxidant defence system, heat shock genes, LTR-type sequences,A-particle sequences. Evidence for a role of poly ADP-ribosylation in gene expression (235, 236, 237) and malignant transformation has recently been obtained(238). Inhibitors of poly ADP-synthesis suppressed (239, 240, 241) or stimulated (242,-243) transformation in different systems.

Acknowledgments:

During the writing of this article I have profited from discussions with Drs D. Borg, I. Emerit, P. Hornsby, W. Kozumbo, C. Richter and J. Seegmiller. Original work reported here was supported by the Swiss National Science Foundation and the Swiss Association of Cigarette Manufacturers.

REFERENCES
1. Totter JR: Spontaneous cancer and its possible relationship to oxygen metabolism. Proc Natl Acad Sci USA (77): 1763-1767, 1980.
2. Heston W, Pratt A: Effect of concentration of oxygen on occurrence of pulmonary tumors in strain A mice. J Natl Cancer Inst (22):707-711, 1959.
3. Sanford K, Parshad R, Jones G, Handleman S, Garrison C, Price F: Role of photosensitization and oxygen in chromosome stability and spontaneous malignant transformation in culture. J Natl Cancer Inst (63):1245-1252, 1979.
4. Zimmerman R, Cerutti P.: Active oxygen acts as a promotor of transformation in mouse embryo fibroblasts C3H/10T1/2/C18. Proc Natl Acad Sci (81):000, 1984, (April)
5. Van Duuren B, Nelson N, Orros L.: Carcinogenicity of epoxides, lactones and peroxy compounds. J Natl Cancer Inst (31):41-55, 1963.
6. Bock F, Myers H, Fox H: Tumor promoting activity of compounds of interest in the laboratory. Proc Am Assoc Cancer Res (7):7, 1966.
7. Slaga T, Klein-Szanto A, Triplett L, Yotti L, Trosko J: Skin tumor promoting activity of benzoyl peroxide, a widely used free radical generating compound. Science (13):1023-1025, 1981.
8. Slaga T, Solanki V, Logani M: Studies on the mechanism of action of antitumor promoting agents: suggestive evidence for the involvement of free radicals in promotion. In: Nygaard OF, Simic MG (eds) Radioprotectors and Anticarcinogens. Academic Press, New York, 1983, pp 471-485
9. German J: Genes which increase chromosomal instability in somatic cells and predispose to cancer. In: Steinberg A, Bearn A (eds) Medical Genetics. Grune and Stratton, New York. 1972, Vol. 8, pp 61-101.
10. Cerutti P: Abnormal oxygen metabolism in Bloom Syndrome ? In: Natarajan A, Altman J, Obe G (eds) Progress in Mutation Research. Elsevier Biomedical Press, Amsterdam, 1982, Vol. 4, pp 203-214.
11. Chaganti R, Schonberg S, German J: A manifold increase in sister chromatid exchanges in Bloom's Syndrome lymphocytes. Proc Natl Acad Sci USA (71):4508-4512, 1974.
12. Warren S, Trosko J:Elevated spontaenous mutation rate in Bloom syndrome fibroblasts. Proc Natl Acad Sci (78):3133-3137, 1981.
13. Vijayalaxmi H, Evans H, Ray J, German J: Bloom's Syndrome: Evidence for an increased mutation frequency in vivo. Science (221):851-853, 1983.
14. Kennedy A, Radner B, Nagasawa H: Protease inhibitors suppress spontaneously occurring chromosome abnormalities in cells from patients with Bloom's Syndrome. Proc Natl Acad Sci (81):000, in press.
15. Goldstein B, Witz G, Amoruso M, Troll W: Protease inhibitors antagonize the activation of polymorphonuclear leukocyte oxygen consumption. Biochem Biophys Res Commun (88):854-860, 1979.
16. Kennedy A, Little J: Protease inhibitors suppress radiation-induced malignant transformation in vitro.Nature (276):825-826, 1978.
17. Troll W, Klassen A, Janoff A: Tumorigenesis in mouse skin:Inhibition by synthetic inhibitors of proteases.Science (169):1211-1213, 1970.

18. Taylor A, Harnden D, Arlett C, Harcourt S, Lehman A, Stevens S, Bridges B: A human mutation with abnormal radiation sensitivity. Nature (258):427-429, 1975.
19. Taylor A, Rosney C and Campbell J: Unusual sensitivity of Ataxia Telangiectasia cells to bleomycin. Cancer Res (39):1046-1050, 1979.
20. Shiloh Y, Tabor E, Becker J: Cellular hypersensitivity to neocarcinostatin in AT skin fibroblasts. Cancer Res (42):2247- 2249, 1982.
21. Fujiwara Y, Tatsumi M: Repair of mitomycin C damage in mammalian cells and its impairment in FA cells. Biochem Biophys Res Commun (66):592-598, 1975.
22. Nagasawa H, Little J: Suppression of cytotoxic effect of mitomycin C by superoxide dismutase in Fanconi's anemia and dyskeratosis congenita fibroblasts. Carcinogenesis (4):795-799, 1983.
23. Gruenert D, Cleaver J: Repair of psoralen-induced cross-links and monoadducts in normal and repair-deficient human fibroblasts. Mutat.Res. DNA repair reports, in press.
24. DeMol, N., Beijersbergen van Henegouwen, G., and van Beele,B. : Singlet oxygen formation by sensitization of furocoumarins complexed with or bound covalently to DNA. Photochem. Photobiol. 34, 661-671 , 1981.
25. Zbinden I, Cerutti P: Near-ultraviolet sensitivity of skin fibroblasts of patients with Bloom's syndrome. Biochem Biophys Res Commun 98, 579-587, 1981.
26. Hirschi M, Netrawali M, Remsen J, Cerutti P: Formation of DNA single-strand breaks by near-ultraviolet and -rays in normal and Bloom's syndrome skin fibroblasts. Cancer Res. (41)2003-2007, 1981.
27. Joenje H, Arwer F, Eriksson AW, de Koning H, Oostra AB: Oxygen-dependence of chromosomal aberrations in Fanconi's Anemia. Nature (290):142-143, 1981.
28. Shaham M, Becker Y, Cohen M: A diffusable clastogenic factor in Ataxia telangiectasia. Cytogenet Cell Genet (27):1-7, 1980.
29. Emerit I, Jalbert P, Cerutti P: Chromosome breakage factor in the plasma of two Bloom Syndrome patients. Human Genet (61):65-67, 1982.
30. Emerit I, Cerutti P: Clastogenic activity from Bloom's Syndrome fibroblast cultures. Proc Natl Acad Sci USA (78):1868-1872, 1981.
31. Cerutti P, Emerit I, Amstad P: Membrane-mediated chromosomal damage. In: Weinstein IB, Vogel H (eds) Genes and Proteins in Oncogenesis. Academic Press, New York, pp 55-69, 1983.
32. "Radioprotectors and Anticarcinogens" Nygaard OF, Simic MG (eds) Academic Press, New York, 1983.
33. "Protective Agents in Cancer" McBrien DCH, Slater TF (eds) Academic Press, New York, 1983.
34. Modulation and Mediation of Cancer by Vitamines (eds. F. Meysken and K. Prasad). S. Karger, Basel 1983
35. Shamberger RJ, Willis C: Selenium distribution and human cancer mortality. Crit Rev Clin Chem Sci (2):211-221, 1971.
36. Flohé L: Glutathione peroxidase brought into focus. In: Pryor W (ed) Free Radicals in Biology, Academic Press, New York, Vol V, pp 223-254,1981.

37. Shamberger R: Relationship of selenium to cancer I. Inhibition by selenium of carcinogenesis. J Natl Cancer Inst (44):931-935, 1970.
38. Clayton C, Baumann C:Diet and Azo Dye tumors : Effect of diet during a period when the dye is not fed. Cancer Res (9):575-582, 1949.
39. Slaga T, Bracken W: The effects of antioxidants on skin tumor initiation and arylhydrocarbon hydroxylase. Cancer Res (37):1631-1635, 1977.
40. Benedict W, Wheatley W, Jones P: Differences in anchorage-dependent growth and tumorigenesis between transformed C3H/10T1/2 cells with morphologies that are not reverted to a normal phenotype by ascorbic acid. Cancer Res (42):1041-1045, 1982.
41. Prasad K, Edwards-Prasad: Effects of tocopherol (Vitamin E) acid succinate on morphological alterations and growth inhibition in melanoma cells in culture. Cancer Res.(42):550-555, 1982.
42. Kensler T, Bush D and Kozumbo W: Inhibition of tumor promotion by a biomimetic superoxide dismutase. Science (221):75-77, 1983.
43. Borek C, and Troll W: Modifiers of free radicals inhibit in vitro the oncogenic actions of X-rays, bleomycin and the tumor promotor 12-0-tetradecanoylphorbol-13-acetate. Proc. Natl. Acad. Sci.(80): 1304-1307, 1983.
44. Kennedy B, Troll W and Little J: personal communication.
45. Ames B, Cathcart R, Schwiers E, Hochstein P: Uric acid provides an antioxidant defense in humans against oxidant- and radical-caused aging and cancer: A hypothesis. Proc Natl Acad Sci (78):6858-6862, 1981.
46. Fridovich I: The biology of oxygen radicals. Science (201):875-880, 1978.
47. Fridovich I: Oxygen is toxic. Bio Science (27):462-466, 1977.
48. for reviews see in "Free Radicals in Biology", Pryor W (ed) Academic Press, New York, Vol I-VI.
49. Winter P, Smith G: The toxicity of oxygen. Anesthesiology (37): 210-241, 1972.
50. Balin A, Goodman D, Rasmussen H, Cristofalo V: The effect of oxygen tension on the growth and metabolism of WI-38 cells. J Cell Physiol (89):235-250, 1976.
51. Balin A, Goodman D, Rasmussen H and Cristofalo V: Oxygen-sensitive stages of the cell cycle of human diploid cells. J. Cell. Biol. (78):390-400, 1978.
52. Sturrock J, Nunn J: Chromosomal damage and mutations after exposure of chinese hamster cells to high concentrations of oxygen. Mutat. Res (57):27-33, 1978.
53. Yost F.J, Fridovich I: Superoxide and hydrogen peroxide in oxygen damage. Arch. Biochem. Biophys.(175):514-519, 1976.
54. Bruyninckx W, Mason H, Morse S: Are physiological oxygen concentrations mutagenic ? Nature (274): 606-607, 1978.
55. Moody C, Hassan H: Mutagenicity of oxygen free radicals. Proc Natl Acad Sci(79):2855-2859, 1982.
56. Cunningham M, Lokesh B: Superoxide anion generated by potassium superoxide is cytotoxic and mutagenic to chinese hamster ovary cells. Mutat. Res.(121): 299-304, 1983.
57. Emerit I, Keck M, Levy A, Feingold J, Michelson A.M: Activated oxygen species at the origin of chromosome breakage and sister-chromatid exchanges. Mutat. Res.(103): 165-172, 1982.

58. Levin D, Hollstein M, Christman M, Schwiers E, Ames B: A new Salmonella tester strain (TA 102) with AT base pairs at the site of mutation detects oxidative mutagens. Proc Natl Acad Sci (79): 7445-7449, 1982.
59. Bradley M, Erickson L: Comparisons of the effects of hydrogen peroxide and x-irradiation on toxicity, mutation and DNA damage repair in mammalian cells (V-79). Biochim. Biophys. Acta (654): 135-141, 1981.
60. Johansen I, Howard-Flanders P: Macromolecular repair and free radical scavenging in the protection of bacteria against X-rays. Radiat. Res.(24):184-200, 1965.
61. Roots R, Okada S: Protection of DNA molecules of cultured mammalian cells from radiation induced single-strand scission by various alcohols and SH compounds. Int. J. Radiat. Biol.(21): 329-342,1972.
62. RotiRoti J, and Cerutti P: Gamma-ray induced thymine damage in mammalian cells. Int. J. Radiat. Biol.(25): 413-417,1974.
63. Remsen J, RotiRoti J: Formation of 5,6-dihydroxydihydrothymine-type products in DNA by hydroxyl radicals. Int. J. Radiat. Biol. (32):191-194, 1977.
64. Cerutti P, Netrawali M: Formation and repair of DNA damage induced by indirect action of ultraviolet light in normal and Xeroderma Pigmentosum skin fibroblasts, in : Proceedings of VIth Int. Congr. of Radiat. Res. Tokyo (S. Okada, M. Inamura, T. Terashima and H. Yamaguchi etc) pp 423-432, Toppan Printing Co, Ltd, Tokyo, 1979.
65. Zelle B, Reynolds R, Kottenhagen M, Schuite A, Lohman P: The influence of the wavelenghts of ultraviolet radiation on survival, mutation induction and DNA repair in irradiated chinese hamster cells. Mutat. Res.(72): 491-509, 1980.
66. Jacobson E, Krell K, Dempsey M: The wavelength dependence of ultraviolet light-induced cell Killing and mutagenesis in L5178Y mouse lymphoma cells. Photochem. Photobiol.(33): 257-260, 1981.
67. Suzuki F, Hay A, Lankas G, Utsumi H, Elkind M: Spectral dependencies of killing, mutation and transformation in mammalian cells and their relevance to hazards caused by solar ultraviolet radiation. Cancer Res.(41): 4916-4924, 1981.
68. Staberg B, Wulf H, Poulsen T, Klemp P, Brodthagen H: Carcinogenic effect of artificial sunlight and UV-A irradiation in hairless mice. Arch. Dermatol.(119): 641-643, 1983.
69. Causes and Effects of stratospheric ozone reduction : an update. National Academy Press, Washington D.C. 1982.
70. Foote Ch: Photosensitized oxidation and Singlet oxygen : consequences in biological systems. In: Pryor W(ed) "Free radicals in Biology",Academic Press, N.Y. 1976,Vol.II,pp 85-133.
71. Cerutti P: Effects of ionizing radiation on mammalian cells. Naturwissenschaft,(61): 51-59, 1974.
72. For reviews, see series Adv. in Radiat. Biol. (Ed J. Lett and H. Adler) Academic Press, N.Y.
73. Mason R: Free radical metabolites and toxic chemicals. In:Pryor W(ed) "Free Radicals in Biology", Academic Press, New York, 1982, Vol. V, pp 161-222.
74. Borg D, Schaich K: Cytotoxicyty from coupled redox cycling of autoxidizing xenobiotics and metals. Israel J. Chem (23): 1983, in press.

75. Bachur N, Gordon S, Gee M: A general mechanism for microsomal activation of quinone anticancer agents to free radicals. Cancer Res.(38): 1745-1750, 1978.
76. Bellomo G, Jewell S, Thor H, Orrenius S: Regulation of intracellular calcium compartmentation : Studies with isolated hepatocytes and t-butylhydroperoxide. Proc Natl Acad Sci (79): 6842-6846, 1982.
77. Lehninger A, Vercesi A, Bababunmi E: Regulation of Ca^+ release from mitochondria by oxidation-reduction state of pyridine nucleotides. Proc Natl Acad Sci,(75): 1690-1694, 1978.
78. Jewell S, Bellomo G, Thor H, Orrenius S, Smith M: Bleb formation in hepatocytes during drug metabolism is caused by disturbances in thiol and calcium ion homeostasis. Science (217): 1257-1259, 1982.
79. Goodman J, Hochstein P: Generation of free radicals and lipid peroxidation by redox cycling of adriamycine and daunomycin. Biochem. Biophys. Res. Comm.(77): 797-803, 1977.
80. Kimball J, Gleiser C, Wang Y: Vitamin E modification of daunomycin-induced carcinogenesis. Clin. Res. 27, 825A 1979
81. Krywanska E, Piekarski L: Neoplasma (24) 395 1977
82. Biaglow J, Jacobson B, Nygaard D: Metabolic reduction of 4-nitroquinoline-N-oxide and other radical-producing drugs to oxygen-reactive intermediates. Cancer Res.(37): 3306-3313, 1977.
83. Varnes M, Biaglow J: Interactions of the carcinogen 4-nitroquinoline-1-oxide with the non-protein thiols of mammalian cells. Cancer Res.(39): 2960-2965, 1979.
84. Biaglow J, Varnes M, Astor M, Mitchell J, Russo A: Intracellular thiols : Involvement in drug metabolism and radiation response.In: Nygaard C, Simic M(eds) "Radioprotectors and Anticarcinogens", Academic Press, N.Y., 1983, pp 203-236.
85. Ikenaga M, and Kakunaga T: Excision of 4-nitroquinoline-1-oxide damage and transformation in mouse cells. Cancer Res.(37): 3672-3678, 1977.
86. Chabner B: "Bleomycin" In: "Pharmacologic Principles of Cancer Treatment".W.B.Saunders Company, 1982,pp 377-386.
87. Morgan A, Cone R, Elgert T: The mechanism of DNA strand breakage by Vitamin C and superoxide and the protective roles of catalase and superoxide dismutase. Nucl. Acid. Res.(3): 1139-1149, 1976.
88. Ide M, Kaneko M, Cerutti P: Benzo(a)pyrene and ascorbate-$CuSO_4$ induce DNA damage in human cells by indirect action". In: Mc Brien D. and Slater T.(eds) Protective Agents in Cancer, Academic Press, N.Y. 1983, pp 125-140.
89. Mason R, Peterson F, Holtzman :Inhibition of azoreductase by oxygen : the role of the azo anion free radical metabolite in the reduction of oxygen to superoxide. Mol. Pharmacol.(14):665- 1978.
90. Misra H, Fridovich I: The oxidation of phenylhydrazine: Superoxide and mechanism. Biochemistry,(15): 681-687, 1976.
91. Hassan H, Fridovich I: Intracellular production of superoxide radical and of hydrogen peroxide by redox active compounds. Arch. Biochem. Biophys.(196): 385-395, 1979.
92. Valuenzuela A, Rios H, Neiman G: Evidence that superoxide radicals are involved in the hemolytic mechanism of phenylhydrazine. Experientia (33):962-963, 1977.
93. Goldberg B, Stern A: The mechanism of oxidative hemolysis produced by phenylhydrazine. Mol. Pharmacol.(13): 832-839, 1977.

94. Slater T.F: Activation of carbon tetrachloride : chemical principles and biological significance. In:"Free Radicals, lipid peroxidation and cancer" (Eds D. Mc Brien and T.Slater) Academic Press, N.Y. 1982, pp 243-274.
95. National Research Council of Canada (1981) "Polylchlorinated dibenzo(p)dioxins. Criteria for their effects on man and his environment. NRCC Publ. No 18574, Natl Res. Counc. Can, Ottawa.
96. Cole L, Nowell P : Accelerated induction of hepatomas in fast neutron-irradiated mice infected with carbon tetrachloride. Ann. N.Y. Acad. Sci (114):259-267, 1964.
97. Pitot H, Goldsworthy T, Campbell H, Poland A: Quantitative evaluation of the promotion by 2,3,7,8-tetrachlorodibenzo-p-dioxin of hepatocarcinogenesis from diethylnitrosamine.Cancer Res.(40):-3616-3620, 1980.
98. Poland A, Knutson J, Glover E, Kende A: Tumor promotion in the skin of hairlesss mice by halogenated aromatic hydrocarbons. In: Weinstein I.B. and Vogel H (eds) "Genes and Proteins in Oncogenesis", Academic Press, N.Y, 1983, pp 143-161.
99. Stohs S, Hassan M, Murray W: Lipid peroxidation as a possible cause of TCDD toxicity. Biochem. Biophys. Res. Comm.(III): 854-859, 1983.
100. Turrens J, Boveris A: Generation of superoxide anion by the NADH dehydrogenase of bovine heart mitochondria. Biochem. J.(191): 421-427, 1980.
101. Mannering G, In:"Concepts in Drug Metabolism" . Jenner P and Testa B(eds), Marcel Dekker, N.Y., 1981, pp 53-166.
102. Estabrook R, Werringloer J: Tissue hypoxia and ischemia. In: Reivich M. et al.(eds) Advances in Expt. Med. Biol. Plenum Press, N.Y.,1976, pp 19-35.
103. O'Brien, P: Hydroperoxides and superoxides in microsomal oxidations. Pharmacol. Ther.(A 2): 517-536, 1978.
104. Plaa G, Witschi H: Chemicals, Drugs and Lipid Peroxidation.Ann. Rev. Pharmacol. Toxicol.(16): 125-141, 1976.
105. Gunsalus I, Sligar S: Enzymol. Relat. Areas Mol. Biol. 48, 33-203, 1978.
106. Gosalvez M: Carcinogenesis with the insecticide rotenone. Minireview. Life Sciences, (32): 809-816, 1983.
107. Pitot H, Sirica A: The stages of initiation and promotion in hepatocarcinogenesis. Biochem. Biophys. Acta (605): 191-195, 1980.
108. Reddy J.K, Warren J.R, Reddy M.K, Lalwani N.D: Hepatic and renal effects of peroxisome proliferators : biological implications. Ann. N.Y. Acad. Sci(386): 81-110, 1982.
109. Jones D, Eklow L, Thor H, Orrenius S: Metabolism of hydrogen peroxide in isolated hepatocytes : relative contributions of catalase and glutathione peroxidase in decomposition of endogenously generated H_2O_2. Arch. Biochem. Biophys.(210): 505-516, 1981.
110. Chance B, Sies H, Boveris A: Hydroperoxide metabolism in mammalian organs. Physiol. Rev.(59):527-605, 1979.
111. Reddy J, Azarnoff D, Hignite C: Hypolipidaemic hepatic peroxisome-proliferators from a novel class of chemical carcinogens. Nature (283): 397-398, 1980.
112. Ward J, Rice J, Creasia D, Lynch P, Riggs C: Dissimilar patterns of promotion by di(2-ethylhexyl)phtalate and phenobarbital of hepatocellular neoplasia initiated by diethylnitrosamine in B6C3F1 mice.Carcinogenesis (4): 1021-1029, 1983.

113. Warren J, Simmon V, Reddy J: Properties of hypolipidemic peroxisome proliferators in the lymphocyte ^{3}H thymidine and Salmonella mutagenesis assays. Cancer Res.(40): 36-41, 1980.
114. Cohen G, Hochstein P: Glutathione peroxidase : the primary agent for the elimination of hydrogen peroxide in erythrocytes. Biochemistry,(2): 1420-1428, 1963.
115. Cohen G, Hochstein P: Generation of hydrogen peroxide in erythrocytes by hemolytic agents. Biochemistry(3): 895-900, 1964.
116. Goldstein B, Rozen M, Quintavalla J, Amoruso M: Decrease in mouse lung and liver glutatione peroxidase activity and potentiation of the lethal effects of ozone and paraquat by the superoxide dismutase inhibitor diethyldithiocarbamate. Biochem.Pharmacol.(28) 27-30, 1979.
117. Sinet P, Garber P, Jerome H: H_2O_2 production, modification of the glutathione status and methemoglobin formation in red blood cells exposed to diethyldithiocarbamate in vitro. Biochem.Pharmacol. (31): 521-525, 1982.
118. Westman G, Marklund S: Diethyldithiocarbamate, a superoxide dismutase inhibitor, decreases the radioresistance of chinese hamster cells. Radiat. Res. (83):303-311, 1980.
119. Frank L, Wood D, Roberts R: Effect of diethyldithiocarbamate on oxygen toxicity and lung enzyme activity in immature and adult rats. Biochem. Paarmacol. 27, 251-1978.
120. Rannug A, Rannug U: The role of enzyme inhibition in the mutagenicity of dithiocarbamic acid derivatives towards Salmonella typhimurium. Chem. Biol. Interact. 1983.
121. Solanki V, Rana R, Slaga T: Diminution of mouse epidermal superoxide dismutase and catalase activities by tumor promotors. Carinogenesis (2):1141-1146, 1982.
122. Oberley L, Buettner G: Role of superoxide dismutase in cancer : a review. Cancer Res. (39):1141-1149, 1979.
123. Meister A: Glutathione Metabolism and Transport. In Nygaard OF and Simic MG(eds) "Radioprotectors and anticarcinogens. Academic Press, N.Y. 1983, pp 121-151.
124. Wendel A, Feuerstein S: Drug induced lipid peroxidation in mice I: modulation by monooxygenase activity, glutathione and selenium status. Biochem. Pharmacol.(30):2513-2520, 1981.
125. Novogodsky A, Nehring R, Meister A:Inhibition of amino acid transport into lymphoid cells by the glutamine analog L-2-amino-4-oxo-5-chloropentaonate. Proc. Natl Acad Sci(76):4932-4935,1979.
126. Arrick B, Nathan C, Griffith O, Cohn Z: Glutathione depletion sensitizes tumor cells to oxidative cytolysis. J. Biol. Chem. (257): 1231-1237, 1982.
127. Dethmers J, Meister A: Glutathione export by human lymphoid cells: depletion of glutathione by inhibition of its synthesis decreases export and increases sensitivity to irradiation. Proc. Natl Acad Sci (78): 7492-7496, 1982.
128. Brune K, Glatt M, Kälin H, Feskar B : Pharmacological control of prostaglandin and thromboxane release from macrophages. Nature (274):261-263, 1978.
129. Levine L, Ohuchi K : stimulation by carcinogens and promotors of prostaglandin production by dog kidney (MDCK) cells in culture. Cancer Res. (38) :4142-4146, 1978.

130. Bresnick E, Meunier P, Lamden M: Epidermal prostaglandins after topical application of a tumor promotor. Cancer Letters(7): 121-125, 1979.
131. Wertz P, Muller G: Rapid stimulation of phospholipid metabolism in bovin lymphocytes by tumor promoting phorbol esters. Cancer Res. (38): 2900-2904, 1978.
132. Marks F, Fürstenberger G, Kownatzki E: Prostaglandin E-mediated mutagenic stimulation of mouse epidermis in vivo by divalent cation ionophore A 23187 and by tumor promotor 12-O-tetradecanoyl-phorbol-13-acetate. Cancer Res.(41): 696-702, 1981.
133. Troll W, Witz G, Goldstein B, Stone D, Sugimura T: The role of free oxygen radicals in tumor promotion and carcinogenesis. In : Hecker E, Fusenig W, Kunz W, Marks F and Thielmann H. (eds) Carcinogenesis and Biological Effects of Tumor Promotors. Carcinogenesis vol. 7, Raven Press, New-York, 1982, pp 593-597.
134. Cerutti P, Remsen J: Formation and repair of DNA damage induced by oxygen radical species in human cells. In:Nichols W and Murphy D (eds). DNA repair Processes. Miami : symposia specialists, 1977, pp 147-166.
135. Ivanovic V, Weinstein I.B. : Glucocorticoids and benzo(a)pyrene have opposing effects on EGF receptor binding. Nature (293):404-406, 1981.
136. Peterhans E, Albrecht H, Wyler R : Detection of H-2 and Sendai virus antigens by chemiluminescence. J. Immunol. Methods (4): 295-302, 1981.
137. Morley J, Bray M.A, Jones R.W, Nugteren D.H, van Dorp D.A: Prostaglandin and thromboxane production by human and guinea-pig macrophages and leucocytes. Prostaglandins (17) : 729-736, 1979.
138. Humes J, Burger S, Galavage M, Kuehl F.A. Jr, Wightman P.D, Dahlgren M.E, Davies P, Bonney R.J: The diminished production of arachidonic acid oxygenation products by elicited mouse peritoneal macrophages :possible mechanisms. J Immunol (124) : 2110-2116, 1980.
139. Peterhans E, Albrecht H, Wyler R: Detection of H-2 and Sendai virus antigens by chemiluminescence. J. Immunol. Methods (4):295-302, 1981.
140. Hafeman D, Lucas Z: Polymorphonuclear leukocyte-mediated antibody-dependent, cellular toxicity, against tumor cells . Dependence on oxygen and the respiratory burst. J. Immunol.(123): 55-62, 1979.
141. Hatefi Y, Hanstein W: lipid oxidation in biological membranes. I: lipid oxidation in submitochondrial particles and microsomes induced by chaotropic agents. Arch. Biochem. Biophys.(138): 73-86, 1970.
142. Cerutti P, Amstad P, Emerit I: Tumor promotor phorbol-myristate-acetate induces membrane-mediated chromosomal damage. In:Nygaard OF and Simic MG (eds) "Radioprotectors and Anticarcinogens" Academic Press, N.Y., 1983, pp 527-538.
143. Egan R, Paxton J, Kuehl F.A: Mechanism for irreversible self-deactivation of prostaglandin synthetase. J. Biol. Chem.(251):7329-7335, 1976.
144. Rahimtula A, O'Brien P.J: The possible involvement of singlet oxygen in prostaglandin biosynthesis. Biochem. Biophys. Res. Comm.-(70): 893-899, 1976.

145. Sugioka K, Nakano M: A possible mechanism of the generation of singlet molecular oxygen in NADH-dependent microsomal lipid peroxidation. Biochem. Biophys. Acta,(423): 203-216, 1976.
146. Van Furth R: Mononuclear phagocytes : functional aspects. Martins Nijkoff Publishers. The Hague, 1980.
147. Roder J, Pross H: The biology of the human natural Killer cell. J. Clin. Immunol.(2): 249-263, 1982.
148. Castagna M, Takai Y, Kaibuchi K, Sano K, Kikkawa U, Nishizuka Y: Direct activation of calcium-activated phospholipid dependent protein kinase by tumor-promoting phorbol esters. J. Biol. Chem. (257):1847-1851, 1982.
149. Niedel J, Kuhn J, Vandenbark E: Phorbol diester receptor co-purifies with protein kinase C. Proc Natl Acad Sci(80): 36-40, 1983.
150. Ashendel C, Staller J, Boutwell R: Protein kinase activity associated with a phorbol ester receptor purified from mouse brain. Cancer Res.(43): 4333-4337, 1983.
151. Emerit I, Cerutti P: Tumor promotor phorbol-12-myristate-13-acetate induces a clastogenic factor in human lymphocytes. Proc Natl Acad Sci(79): 7509-7513, 1982.
152. Mead J: Free radical mechanisms of lipid damage and consequences for cellular membranes.I :Pryor W.(ed)"Free Radicals in Biology", Academic Press, N.Y. 1976, vol 1, pp 51-68.
153. Esterbauer H: Aldehydic Products of lipid peroxidation.In:Mc Brien D, and Slater T (eds)"Free Radicals, lipid peroxidation and Cancer". Academic Press, N.Y. 1982, pp 101-128.
154. Hornsby P, Crivello J: The role of lipid peroxidation and biological antioxidants in the function of the adrenal cortex. Part. I : A background review. Molecular and Cellular Endocrinology (30): 1-20, 1983.
155. Tappel A: Measurement of and protection from in vivo lipid peroxidation.In:Pryor W (ed)"Free Radicals in Biology", Academic Press, N.Y., 1980, vol. IV, pp 2-47.
156. Borg D, Schaich K, Elmore J, Bell J: Cytotoxic reactions of free radical species of oxygen.Photochem. Photobiol.(28): 887-907,1978.
157. O'Brien R, Parker J: Oxidation induced lymphocyte transformation. Cell (7): 13-20 (1976).
158. Srinivas L, Coburn N: personal communication.
159. Fridovich I: Oxygen radicals, hydrogen peroxide and oxygen toxicity.In:Pryor W (ed)"Free Radicals in Biology".Academic Press, N.Y. 1976, vol I, pp 239-277.
160. Raleigh J, Shum F: Radioprotection in model lipid membranes by hydroxyl radical scavengers : Supplementary role for α-tocopherol in scavenging secondary peroxy radicals.In:Pryor W (ed)"Free Radicals in Biology". Academic Press, N.Y.vol. IV, pp 87-102 1980.
161. Witting A: Vitamin E and lipid antioxidants in free radical initiated reactions.In:Pryor W (ed)"Free Radicals in Biology, Academic Press, N.Y., 1980,vol IV, pp 295-320.
162. Csallany A.S, Draper H.H: Determination of N,N'-diphenyl-p-phenylene diamine in animal tissues.Proc. Soc. Exp. Biol. Med.(104):-739-742, 1960.
163. Neumann N: Methods in Enzymology (25):393-400, 1972
164. Fliss H, Weissbach H, Brot N: Oxidation of methionine residues in proteins of activaated human neutrophils. Proc. Natl. Acad. Sci, (80):7160-7164, 1983

165. Jeffrey E, Nerland D, El-Azhary R, Mannering G:Microsomes and Drug Oxidations. In:Ullrich V, Roots A, Hildebrandt R, Estabrook R, Conney A (eds). Pergamon Press, Oxford, 1977, pp 323-330.
166. Kono Y, Fridovich I: Superoxide radical inhibits catalase. J. Biol. Chem. (257):5751-5754, 1982.
167. Christophersen B.O : Biochim. Biophys. Acta (189):387-389,1969.
168. Tyrrell R: Rec A$^+$-dependent synergism between 365 nm and ionizing radiation in log phase Escherichia Coli. A model for oxygen-dependent near-UV inactivation by disruption of DNA repair. Photochem. Photobiol.(23): 13-20, 1976.
169. Grafstrom R, Fornace A, Autrup H, Lechner J, Harris C:Formaldehyde damage to DNA and inhibition of DNA repair in human bronchial cells. Science (220): 216-218, 1983.
170. Miller E.C: Some current perspectives in chemical carcinogenesis and experimental animals : presidential address. Cancer Res.(38): 1479-1496, 1978.
171. Cerutti P: Repairable damage in DNA. In: Hanawalt, P, Friedberg E, Fox C (eds). DNA repair mechanisms, Academic Press, N.Y. 1978, pp 717-722.
172. Pietronigro D, Jones W, Kalty K, Demopoulos H: Interaction of DNA and liposomes as a model for membrane-mediated DNA damage. Nature (267):78-79,1977.
173. Weiss S, Lampert M, Test S: Long-lived oxidants generated by human neutrophils :characterization and bioactivity. Science (222):625-627,1983.
174. Webb R, Photochemical and Photobiological Reviews 2. In:Smith K (ed). Plenum Press, N.Y. 1977, pp 198-201.
175. Arcos J, Argus M.In:Chemical Induction of Cancer, Structural Bases and Biological Mechanisms, Academic Press, N.Y. 1974, vol. IIB, pp 111-121.
176. See in "Prostaglandins and Cancer : First International Conference".In: Powles T et al (eds), Alan R. Liss Inc. New-York, 1982.
177. Hariharan P, Cerutti P: Formation of products of the 5,6-dihydroxy-dihydrothymine type by ultraviolet light in Hela Cells. Biochemistry (16):2791-2795, 1977.
178. Amstad P, Levy A, Emerit I, Cerutti P: Carcinogenesis, in press.
179. Emerit I, Cerutti P: Tumor promotor phorbol-12-myristate-13-acetate induces chromosomal damage via indirect action. Nature (293): 144-146, 1981.
180. Emerit I, Levy A, Cerutti P: Suppression of tumor promoter phorbolmyristate acetate-induced chromosome breakage and sister-chromatid exchanges. Mutat. Res. (103):165-172, 1983.
181. Leadon S, Hanawalt P: Monoclonal antibody to DNA containing thymine glycol. Mutat. Res. (112):191-200, 1983.
182. Kinsella A, Gainer H, Butler J: Investigation of a possible role for superoxide anion production in tumor promotion. Carcinogenesis (4): 717-719, 1983.
183. Lesko S, Lorentzen R, Ts'o P: Role of superoxide in deoxyribonucleic acid strand scission. Biochemistry (19): 3023-3028, 1980.
184. Brawn K, Fridovich I: DNA strand scission by enzymatically generated oxygen radicals. Arch. Biochem. Biophys.(206): 414-419, 1981.
185. Rhaese H, Freese E: Chemical analysis of DNA alterations. 1. Base liberation and backbone breakage of DNA and oligodeoxyadenylic acid induced by hydrogen peroxide and hydroxylamine. Biochim. Biophys. Acta,(155):476-490, 1968.

186. Bautz-Freese, E., Gerson, J., Taber, H., Rhaese, H. and Freese, E. : Inactivating DNA alterations induced by peroxides and peroxide-producing agents. Mutat. Res. 4 : 517-531, 1967.
187. Bradley, M. Hsu, I. and Harris, C. : Relationship between sister chromatid exchange and mutagenicity, toxicity and DNA damage. Nature 282 : 318-320, 1979.
188. Conger, A. and Fairchild, L. : Breakage of chromosomes by oxygen. Proc Natl Acad Sci 38 : 289-299, 1952.
189. Emerit I:Chromosome Breakage factors : Origin and possible significance. In: Natarajan AT et al(eds). Progress in Mutation Research, Elsevier Biomedical Press, Amsterdam, vol.4, pp 61-74.
190. Sasaki M, Matsubara S: Free radical scavenging in protection of human lymphocytes against chromosome aberration formation by gamma-ray irradiation. Int. J. Radiat. Biol.(32): 439-445, 1977.
191. Nordenson I: Effects of superoxide dismutase and catalase on radiation-induced chromosome aberrations : dose and cell cycle dependence. Hereditas (89): 163-167, 1978.
192. Demple B, Halbrook J: Inducible repair of oxidative DNA damage in Escherichia Coli. Nature (304): 466-468, 1983.
193. see e.g. "DNA repair mechanisms" (eds. P. Hanawalt, E. Friedberg and C. Fox) Academic Press, N.Y. 1978.
194. In : "The repair of genomic damage in living tissue" Friedberg E, Bridges B, Fox I (eds), Alan R. Liss, N.Y. 1984, in press.
195. Dzarlieva R, Fusenig N: Tumor promotor 12-O-tetradecanoylphorbol-13-acetate enhances sister chromatid exchanges and numerical and structural chromosome aberrations in primary mouse epidermal cultures. Cancer Letters (16):7-17, 1982.
196. Parry J.M, Parry E.M, Barret J.C: Tumor promoters induce mitotic aneuploidy in yeast. Nature (294): 263-265, 1981.
197. Birnboim H.C: DNA strand breakage in human leukocytes exposed to a tumor promoter, phorbol-myristate-acetate. Science,(215): 1247-1249, 1982.
198. Barsoum J, Varshavsky A: Mitogenic hormones and tumor promotors greatly increase the incidence of colony-forming cells bearing amplified dihydrofolate reductase genes. Proc Natl Acad Sci(80): 5330-5334, 1983.
199. Tlsty T, Brown P, Johnson R, Schimke R: Enhanced frequency of generation of methotrexate resistance and gene amplification in cultured mouse and hamster cell lines. In:Schlimke R (ed)Gene Amplification.Cold Spring Harbor Laboratory, 1982, pp 231-238.
200. Oppenheim J, Fishbein W: Induction of chromosome breaks in cultured normal human leukocytes by potassium arsenite, hydroxyurea and related compounds. Cancer Res.(25): 980-982, 1965.
201. Przybyszewski W, Malec J: Protection against hydroxy-urea-induced cytotoxic effects in L5178Y cells by free radical scavengers. Cancer Letters (17): 223-228, 1982.
202. Makino F, Okada S: Effects of ionizing radiation on DNA replication in cultured mammalian cells. Rad. Res.(62): 37-51, 1975.
203. Painter R,Young B: X-ray induced inhibition of DNA synthesis in chinese hamster ovary-, human Hela and mouse L-cells. Radiat. Res.(64): 648-656, 1975.
204. Kaufmann W, Schwartz J: Inhibition of replicon-initiation by 121-Otetradecanoylphorbol-13-acetate. Biochem. Biophys. Res. Comm.-(103): 82-89, 1981.
205. Seegmiller J, Cerutti P : unpublished.

206. Emerit I, Cerutti P: Clastogenic action of tumor promotor phorbol-12-myristate-13-acetate in mixed human leukocyte cultures. Carcinogenesis(4): 1313-1316, 1983.
207. Bonney R, Humes J: Studies on a destructive oxidant released in the enzymatic reduction of prostaglandin G_2 and other hydroperoxy acids. In: Autor A (ed) Pathology of Oxygen, Academic Press, N.Y. 1982, pp 175-189.
208. Kuehl F, Ham E, Egan R, Dougherty H, Bonney R, Humes J: "Studies on a destructive oxidant released in the enzymatic reduction of prostaglandin G_2 and other hydroperoxy acids".In:Autor A (ed)- Pathology of Oxygen".Academic Press, N.Y., 1982,pp 175-189.
209. Murray H, Cohen Z: Macrophage oxygen-dependent antimicrobial activity. III: Enhanced oxidative metabolism as an expression of macrophage activation. J. Exp. Med.(152): 1596-1609, 1980.
210. Murray H, Inangbhanich Ch, Nathan C, Cohen Z: Macrophage oxygen-dependent antimicrobial activity. II: The role of oxygen intermediates. J. Exp. Med.(150): 950-964, 1979.
211. Ashendel C, Boutwell R: Prostaglandin E and F levels in mouse epidermis are increased by tumor promoting phorbol esters. Biochem. Biophys. Res. Comm. (90):623-627, 1979.
212. Fürstenberger G, Richter H, Fusenig N, Marks F: Arachidonic acid and prostaglandin E_2 release and enhanced cell proliferation induced by the phorbolester TPA in a murine epidermal cell line. Cancer Letters (11): 191-198, 1981.
213. Wrighton S, Pai J.-K, Mueller G: Demonstration of two unique metabolites of arachidonic acid from phorbol ester-stimulated bovine lymphocytes. Carcinogenesis(4): 1247-1251, 1983.
214. Weitzman S, Stossel T: Mutation caused by human phagocytes. Science (212): 546-547, 1981.
215. Weitberg A, Weitzman S, Destrempes M, Latt S, Stossel T: Stimulated human phagocytes produce cytogenetic changes in cultured mammalian cells. New Engl. J. Med.(308): 26-30, 1983.
216. Birnboim H: Importance of DNA strand break damage in tumor promotion.In:Nygaard O and Simic M (eds)"Radioprotectors and Anticarcinogens". Academic Press, N.Y.,1983, pp 539-556.
217. Wang T.V, Cerutti P: Effect of formation and removal of aflatoxin B_1: DNA adducts in 10T1/2 mouse embryo fibroblasts on cell viability. Cancer Res.(40): 2904-2909, 1980.
218. Amstad P, Cerutti P: DNA binding of aflatoxin B_1 by co-oxygenation in mouse embryo fibroblasts C3H/10T1/2, Biochem. Biophys. Res. Comm.(112): 1034-1040, 1983.
219. Wang T. V, Cerutti P: Formation and removal of aflatoxin B_1 induced DNA lesions in epithelioid human lung cells. Cancer Res. (39): 5165-5170, 1979.
220. Cerutti P, Wang V.T, Amstad P: Reactions of aflatoxin B_1 damaged DNA in vitro and in situ in mammalian cells, in Carcinogenesis : Fundamental and Environmental Effects (B. Pullamn, P.O.P. Ts'o and H. Gelboin, eds.) pp 465-477, D. Reidel Publ. Company, 1980.
221. Hiller K, Wilson R: Hydroxyl-free radicals and anti-inflammatory drugs : biological inactivation studies and reaction rate constants. Biochem. Pharmacol.(32): 2109-2111, 1983.
222. Hassid A, Levine L: Induction of fatty acid cyclooxygenase activity in canine kidney cells (MDCK) by benzo(a)pyrene, J. Biol. Chem.(252): 6591-6593, 1977.

223. Slaga T, Fisher S, Nelson K, Gleason G: Studies on the mechanism of skin tumor promotion : evidence for several stages of promotion. Proc. Natl. Acad. Sci(77): 3659-3663, 1980.
224. Fürstenberger G, Sorg B, Marks F: Tumor promotion by phorbolester in skin : evidence for a memory effect. Science (220): 89-91, 1983.
225. Coburn N, Former K, Nelson K, Juspa S: Tumour promotor induces anchorage independence irreversibly.Nature (281): 589-591, 1979.
226. Coburn N: Tumor promotor produces anchorage independence in mouse epidermal cells by an induction mechanism. Carcinogenesis (1) 951-1980.
227. Burns F, Albert R, Altschuler B, Morris E : Approach to risk assessment based on data from the mouse skin initiation promotion model. Environmental Health Perspectives (50):309-320, 1983.
228. Mattern M.R, Hariharan P.V, Cerutti P: Selective excision of gamma-ray damaged thymine from the DNA of cultured mammalian cells, Biochim. Biophys. Acta (395): 48-55, 1975.
229. Ormerod M: "Radiation-induced strand breaks in the DNA of mammalian cells, in "Biology of Radiation Carcinogenesis (Eds J. Juhas, R. Tennant and J. Regan) pp 67-92, Raven Press, N.Y., 1976.
230. Kozumbo W, Seed J, Kensler T: Inhibition by 2(3)-tert-butyl-4-hydroxyanisole and other antioxidants of epidermal ornithine decarboxylase activity induced by 12-O-tetradecanoylphorbol-13-acetate. Cancer Res.(43): 2555-2559, 1983.
231. Friedman J, Cerutti P: The induction of ornithine decarboxylase by phorbol-12-myristate-13-acetate or by serum is inhibited by antioxidants. Carcinogenesis (4): 1425-1427, 1983.
232. Bishop J.M :Cellular oncogens and retroviruses. Ann. Rev. Biochem (52): 301-354, 1983.
233. Bishop J.M: Cancer genes come of age.Cell (32): 1018-1020, 1983.
234. Hayaishi O, Ueda K. eds : "ADP-ribosylation reactions". Biology and Medicine, Academic Press, N.Y. 1982.
235. Johnstone A, Williams G: Role of DNA breaks and ADP-ribosyl transferase activity in eukaryotic differentiation demonstrated in human lymphocytes. Nature (300): 368-370, 1982.
236. Althaus F, Lawrence S, He Y-Z, Sattler G, Tsukada Y,Pitot H: Effects of altered ADP-ribose in metabolism on expression of fetal functions by adult hepatocytes. Nature (300):366-368, 1982.
237. Farzaneh F, Zalin R, Brill D, Shall S: DNA strand breaks and ADP-ribosyl transferase activation during cell differentiation. Nature(300): 362-366,1982.
238. Sugimura T, Miwa N : Poly(ADP-ribose) and cancer research. Carcinogenesis (4):1503-1506, 1983.
239. Milo G: In vitro transformation of human cells : modulation of early gene expression preceding carcinogen-induced events".In:- Harris C, Autrup H (eds). Human Carcinogenesis, Academic Press, N.Y.1983 pp 431-449.
240. Kun E, Kirsten E, Milo G, Kurian P, Kumari H :Cell cycle dependent intervention and in vitro poly(ADP-ribosyl)ation of nuclear proteins in human fibroblasts. Proc Natl Acad Sci (80):7219-7223, 1983.
241. Borek C, Morgan W, Ong A, Cleaver J: Inhibition of malignant transformation in vitro by inhibitors of poly ADP-ribose synthesis. Proc Natl Acad Sci (81):243-247, 1984.

242. Takahashi S, Ohnishi T, Denda A, Konishi Y : Enhancing effect of 3-aminobenzamide on induction of γ-glutamyl transpeptidase positive foci in rat liver. Chem. Biol. Interactions (39):363-368, 1982.
243. Kasid U, Dritschila A, Lubet C, Smulson M : personal communication.

Chapter 7

POSSIBLE INVOLVEMENT OF ARACHIDONATE PRODUCTS IN TUMOR PROMOTER
INHIBITION OF CELL-CELL COMMUNICATION
J.E. Trosko, C. Aylsworth, C. Jone, and C.C. Chang
Michigan State University
East Lansing, MI 48824

                                                                          Page

I.   Role of intercellular communication in the regulation
     of growth and development 170

II. Evidence of tumor promoter-inhibition of intercellular
     communication 172

III. Prostaglandins and tumor promotion. 174

IV. The role of arachidonate metabolites on cell-cell
     communication 178

References. 183

S.M. Fischer and T.J. Slaga (eds.), ARACHIDONIC ACID METABOLISM AND TUMOR
PROMOTION. Copyright © 1985. Martinus Nijhoff Publishing, Boston. All rights reserved.

7. POSSIBLE INVOLVEMENT OF ARACHIDONATE PRODUCTS IN TUMOR PROMOTER INHIBITION OF CELL-CELL COMMUNICATION*

J.E. TROSKO[1], C. AYLSWORTH[2], C. JONE[1], and C.C. CHANG[1]

I. ROLE OF INTERCELLULAR COMMUNICATION IN THE REGULATION OF GROWTH AND DEVELOPMENT

The complex, healthy multicellular organism is the ultimate product of a hierarchy of levels of organization, each interacting with each other and with the environment via different kinds of feedback or communicating mechanisms (1,2). Unless there is a functional hierarchical organization of the molecular, biochemical, cellular, physiological and psychological (in the case of human beings), this state of health cannot be maintained. Proper genetic information in the form of DNA molecules and normal expression of the genetic information are needed to affect the appropriate adaptive biochemistry. Normal regulation of cell proliferation and differentiation is necessary for healthy growth and function of organ systems (3). Proper function of organ systems, in turn, leads to adaptive responses on the whole organism level. Consequently, interference with any of these levels can influence the functions of levels above and below, because of interacting connections between the levels (1,2).

---

*This manuscript was based on research supported by an EPA grant (R808587) and an NCI grant (CA21104) to J.E.T., a NCI Postdoctoral Training Fellowship grant (CA09284) to C.A.

[1]Department of Pediatrics and Human Development and

[2]Department of Anatomy, College of Human Medicine, Michigan State University, East Lansing, Michigan 48824.

A delicate orchestration of intercellular feedback of molecular signals is needed for an organism to regulate whether cells proliferate and differentiate properly (3-6 Potter, 1980; Sachs, 1980; Loewenstein, 1979). The "multi- or pluripotent" stem cell of a given tissue has the ability to proliferate, to make more of itself, as well as to differentiate into a series of highly specialized daughter cells. Proper genetic information within this cell, appropriate expression of that genetic information in the cell to determine if it is to divide or differentiate, functional communication between the stem cells and the differentiated daughter cells, as well as functional communication of this tissue with other tissues in the body, help to maintain a homeostasis (stable equilibrium between different, but interdependent, elements in an organism), which is necessary for health.

Intercellular communication can be classified into two or three (7) generic types. Transfer of molecular signals from cells of one differentiation or tissue type to another type over a distance via an extracellular route (i.e., communication between cells of the immune system; hormone-producing cells with the hormone-receptor target cells, etc.) represents one type. Transfer of relatively small molecular weight, phosphorylated or charged molecules and ions via gap junctions from cells in contiguous contact represents the other type (8-10).

In principle, disruption of these delicate homeostatic feedback systems can be either adaptive or maladaptive, depending on circumstances (11,12). Disruption of intercellular communication can be brought about by at least three fundamental biological processes induced by various physical or chemical agents. These processes are a) mutagenesis (<u>heritable</u> alteration of genetic information); b) gene modulation (altered <u>expression</u> of genetic information; i.e., derepression or repression of genes; altered protein and enzyme function); or c) cytotoxicity (cell death) (13).

In other words, the mutation of genes controlling the synthesis, transfer, reception and translation of molecular signals could permanently inhibit intercellular communication. Reversible modulation of the gene or gene products needed for either of the two types of intercellular communication by exogenous or endogenous chemicals could interfere with intercellular communication. Finally, by the death of either the "signaling" or "receiving" cell, disruption of intercellular communication can result.

II. EVIDENCE OF TUMOR PROMOTER-INHIBITION OF INTERCELLULAR COMMUNICATION

The multistep nature of carcinogenesis has been conceptualized as consisting of initiation, promotion and progression stages (17). The relatively permanent process of initiation refers to the conversion of a single normal cell to a "premalignant" or initiated cell. Promotion appears to be that process which brings about the clonal and selective amplification of the initiated cell, facilitating other genomic changes which ultimately convert an initiated cell to a malignant cell (15,16). Recently, Hennings et al. (17) have data suggesting that malignant conversion during the progression of tumorigenesis appears to be the result of an additional (other than the initial "initiation" event) mutagenic event.

Yotti et al. (18) and Murray and Fitzgerald (19), using two different approaches, showed that the tumor promoting, but not the non-tumor promoting, phorbol esters could inhibit metabolic cooperation (a form of gap junction-mediated intercellular communication). Both groups speculated that by isolating the initiated cell from the mitotic-suppressing influence of the surrounding normal cells, the initiated stem cell had the chance to multiply into a clone of abnormally differentiated cells. Subsequently, using either electrocoupling, dye or radioisotope transfer techniques, or the metabolic cooperation assay in other cell systems, a wide variety of known or suspected tumor promoters has been shown to inhibit intercellular communication at non-cytotoxic levels (20-42).

With the demonstration that certain tumor promoters could inhibit electrocoupling between cells [a direct measure of gap junction communication] (24,25), it was further shown that the tumor promoting, 12-0-tetradecanoyl phorbol 13-acetate but not the non-promoting, 4αPDD, drastically reduced the amount of membrane-associated gap junctions on Chinese hamster V79 cells (43).

Of the many phenotypic alterations which are associated with malignancy, the inability to contact inhibit or to exert a normal ability to perform gap junction-mediated intercellular communication (44,45) is among that list. Since normal cells have the ability to regulate gap junction-mediated intercellular communication, and since many, if not all tumor promoting conditions decrease gap junctional communication, the initiation/promotion/progression model of carcinogenesis might include, among other changes, the dysfunctional control of the gene(s) [oncogenes?] which regulate gap junctions (12,46,47).

Genes controlling gap junction structure and function must be viewed as highly conserved during evolution because of the dependence of important development and differentiation functions on gap junction-mediated intercellular communication (7). Mutation of these genes in the germ line would have created devastating effects on the survival of the species. Mutations of these genes in somatic tissue, depending on circumstances [i.e., early in organogenesis or in initiated stem cell tissue (48)], would affect normal differentiation of tissue. It has been speculated (46) that some proto-oncogenes, which seem to be highly conserved genes and to be important for normal cellular functions (49,50), might include genes regulating gap junctions.

Several observations linking tumor promoting chemicals with viral carcinogenesis have been made. Tumor promoter-induced viral gene expression (51-57), protein-kinase activation by both tumor promoters and certain viral gene products (58-65), as well as altered intercellular communication in viral transformed (66) and tumor promoter treated cells (18-42), suggest a possible common mechanism involving an altered membrane function. The biochemical nature of this speculated altered function, although not known, might involve the arachidonate products in some cells.

III. PROSTAGLANDINS AND TUMOR PROMOTION

The prostaglandins have been implicated as important factors in many physiological processes including cellular growth control mechanisms. Recent interest in the involvement of arachidonate metabolites, especially the prostaglandins, in cancer is based upon the following observations: (a) the increased secretion of prostaglandins by neoplastic tissue (67); (b) the stimulation of prostaglandin biosynthesis by classic tumor promoters (68); (c) the influence of prostaglandins on tumor invasion, metastasis, angiogenesis and osteolysis (69-74); (d) the influence of prostaglandins on cellular differentiation (74); (e) the influence of prostaglandins and prostaglandin synthesis inhibitors on normal and neoplastic cell proliferation *in vitro* and *in vivo* (74); and (f) the influence of prostaglandins and prostaglandin synthesis inhibitors on tumor initiation, promotion and progression (74). However, results concerned with examining the role of prostaglandins and cancer are often confounded by the complexity of prostaglandin biosynthesis and control, the diverse influences of prostaglandin synthesis inhibitors, and the broad range of physiological effects of the different prostaglandins. Many investigations examining the role of prostaglandins and cancer utilize steroidal and nonsteroidal prostaglandin synthesis inhibitors, particularly indomethacin. These prostaglandin synthesis inhibitors, while effective, may have other effects *in vivo* which are not directly related to prostaglandins per se. Furthermore, the diverse and often opposing effects of each specific prostaglandin make interpretation of results, in which prostaglandin synthesis "in general" is inhibited, difficult. The following discussion will attempt to briefly summarize the literature concerned with prostaglandins and tumor promotion.

Elevated prostaglandins of the E_2 and $F_{2\alpha}$ series and increased prostaglandin synthetase activities are associated with a wide range of neoplastic tissues (67,75). Increased prostaglandin (mostly E_2 and $F_{2\alpha}$) tissue and/or plasma levels are found in medullary thyroid tumors (67), human breast cancer (73), human lung carcinomas (76) and others (67). Experimental neoplasms including maloney sarcoma virus-induced mouse tumors (77), transplantable mouse fibrosarcomas (78,79), dimethylbenzanthracene-induced rat mammary tumors (80), VX_2 rabbit carcinomas (81) produce increased levels of prostaglandins, especially E_2 in vivo. However, it remains unclear whether these levels reflect increased levels of prostaglandin biosynthesis in the neoplastic tissue or whether these levels reflect some sort of host response to the tumor tissue (74). Levine (82) reports increased prostaglandin production by mouse fibrosarcoma cells in vitro.

Further evidence implicating prostaglandins in neoplastic processes, i.e., tumor promotion, are the observations by many investigators that tumor promoters, particularly TPA, increase prostaglandin biosynthesis. Tumor promoting phorbol esters such as TPA increase prostaglandin E_2 levels and enhance cell proliferation in mouse epidermal tissue in vivo (83,84) and in canine kidney cells (68), in bone organ cultures (85), in Hela cells (86), in murine epidermal cells (87,88) and others (89) in vitro. Furthermore, inhibitors of prostaglandin biosynthesis, such as indomethacin, block the elevation of prostaglandin biosynthesis, inhibit the induction of ornithine decarboxylase, and the cell proliferative responses to TPA (90,91). The TPA-induced proliferation of a murine cell line is associated with the induction of a cascade of arachidonate metabolism as measured by the release of ^{14}C-arachidonic acid and ^{14}C-prostaglandin E_2 into the culture medium (92). Other tumor promoters, including teleocidin B, carcinogens and growth factors also stimulate prostaglandin synthesis (68,89,93). Therefore, prostaglandins may have an intricate role in mediating many of the biochemical effects of tumor promoters such as TPA.

Reports on the influence(s) of prostaglandins on cell proliferation in vitro are confounding. Prostaglandins of the F and E series stimulate proliferation of some in vitro cell systems (94-96). However, other studies show an inhibitory influence of exogenously added prostaglandins on proliferation of cells in culture (97-100). Furthermore, Santoro et al. (97) have shown that indomethacin enhances B-16 melanoma cell proliferation in culture. Thomas et al. (100) also have reported that indomethacin enhances proliferation of Hep-2 and Hela cells in vitro. Other studies also show that nonsteroidal anti-inflammatory agents, including indomethacin, which inhibit prostaglandin biosynthesis stimulate cell proliferation in vitro (94,101).

Identifying the influence of prostaglandins on cell proliferation in vivo are complicated by the wide range of physiological effects these substances, and their synthesis inhibitors have in various tissues. Indomethacin completely inhibits TPA-induced cell proliferation in normal mouse epidermis when applied 1 hr. prior to TPA treatment (102). In the same report (102), administration of prostaglandin E_2 reversed the inhibition of TPA-induced epidermal hyperplasia by indomethacin, suggesting that prostaglandin E_2 or related compounds may mediate the hyperproliferative effects of TPA. Further data by Furstenberger and Marks (103) suggest that the increase in prostaglandin E_2 synthesis by TPA may be an obligatory event in the induction of epidermal cell proliferation by TPA. Other studies also implicate arachidonic acid and its metabolites in hyperproliferative processes in the skin

(104,105). Prostaglandins also may be involved in the proliferative responses following partial hepatectomy since prostaglandin E equivalents are increased in regenerating liver tissue (106). Finally a prostaglandin $F_{2\alpha}$ analog accelerates jejunal-crypt epithelial cell proliferation in rats (107). The prostaglandin E_2 analog or the cyclo-oxygenase inhibitor (flurbiprofen) are without effect.

Prostaglandins also influence the growth of neoplastic tissues in vivo. Indomethacin, an inhibitor of prostaglandin synthesis, inhibits the growth of HSDM transplantable fibrosarcomas, Moloney sarcoma virus-induced transplantable sarcomas and transplantable methylcholanthrene-induced fibrosarcomas in mice (78,108,109). Indomethacin and acetyl salicylic acid, another inhibitor of prostaglandin biosynthesis, reduce the growth of the Yoshida ascites hepatoma AH130 and prolong the survival time of tumor bearing rats (110). Nonsteroidal anti-inflammatory agents, including indomethacin, inhibit other types of tumor growth and/or increase the survival time of tumor bearing animals (111-118). Furthermore, prostaglandin synthesis inhibitors (eicosatetraynoic acid and indomethacin) prevent the growth promoting effects of high polyunsaturated fat diets on transplantable mammary adenocarcinomas in Balb/c mice (119,120). Flurbiprofen also inhibits the growth of a transplantable syngeneic mouse metastasizing mammary tumor and increases the survival time of tumor bearing animals suggesting an antimetastatic effect of the prostaglandin synthesis inhibitors (121,122). Finally, $PGF_{2\alpha}$ and PGE_2 increase DNA synthesis in mouse transplantable methylcholantrene-induced squamous cell carcinomas and accelerate the evolution of these tumors towards more anaplastic fibrosarcomas (123). Thus, from the aforementioned studies, it would appear that prostaglandins exert a growth enhancing influence on neoplastic tissues in vivo.

However, other reports demonstrate that inhibitors of prostaglandin biosynthesis, such as indomethacin, enhance the growth of the Morris hepatoma No. 7777 in Buffalo rats (116). Also, analogs of prostaglandins E_2 and $F_{2\alpha}$ inhibit human colorectal tumors propagated in immuno-suppressed mice (107). In the same study prostaglandin $F_{2\alpha}$ also inhibited the growth of established DMH-induced rat colonic tumors whereas flurbiprofen accelerated cell division in these tumors (107). Also, prostaglandin $F_{2\alpha}$ inhibits the growth of a hormone-dependent transplantable rat mammary tumor (124). However, this effect was associated with a decrease in serum progesterone concentration which undoubtedly influenced tumor growth. Thus, while it appears that the growth of most tumors are stimulated by prostaglandins, and inhibited by prostaglandin synthesis inhibitors, the response and dependence of any specific type of tumor to/on prostaglandins for growth varies.

IV. THE ROLE OF ARACHIDONATE METABOLITES ON CELL-CELL COMMUNICATION

Given the possible role of arachidonate metabolites, especially the prostaglandins, in experimental tumor promotion and the inhibitory effects of many tumor promoters in gap junction mediated intercellular communication, it seems reasonable to hypothesize that these substances might influence metabolic cooperation.

Using the in vitro V79 Chinese hamster metabolic cooperation assay (137), we have examined the effects of various unsaturated fatty acids on cell-cell communication. Linoleic acid, linolenic acid and arachidonic acid all effectively inhibit metabolic cooperation by V79 cells (Tables 1 and 2; ref 138) at non-cytotoxic concentrations. Since these lipids are potential precursors in the prostaglandin biosynthetic pathway, it is possible that prostaglandins may mediate, at least in part, the inhibitory effects of these fatty acids on metabolic cooperation.

Table 1. Recovery of 6TG^r cells in the presence of TPA, oleic, linoleic and linolenic acids.[a]

Compound	Concentration (µg/ml)	Recovery of 6TG^r cells	% Recovery[b]	Survival[c]
--	--	19.1 ± 2.8	24.0 ± 3.5	--
TPA	0.002	78.0 ± 3.2[d]	98.0 ± 4.0[d]	--
oleic acid	11.9	54.1 ± 1.5[d]	68.0 ± 1.8[d]	97.2 ± 13.5
linoleic acid	12.0	45.1 ± 2.0[d]	56.7 ± 2.5[d]	100.9 ± 4.1
linolenic acid	12.2	65.8 ± 2.8[d]	82.7 ± 3.4[d]	96.8 ± 10.1

[a.] Aylsworth et al. (see ref. 137)
[b.] Metabolic cooperation: % recovery of 100 6TG^r (HG-PRT$^-$) cells co-cultured with 4×10^5 V79 (HG-PRT$^+$) cells relative to the recovery of untreated 6TG^r (HG-PRT$^-$) cells cultured in the absence of V79 (HG-PRT$^+$) cells. (i.e., plating efficiency)
[c.] Cytotoxicity: % Recovery of 6TG^r (HG-PRT$^-$) cells relative to the recovery of untreated 6TG^r (HG-PRT$^-$) cells cultured alone (i.e. plating efficiency)
[d.] $p < 0.001$ vs. solvent control

However, other unsaturated fatty acids, such as oleic acid and palmitoleic acid, also effectively inhibit metabolic cooperation in V79 cells (Tables 1 and 3; refs. 138,139) and are not classically thought to be involved in prostaglandin biosynthesis. Therefore, mechanisms other than via the prostaglandins must also be involved in the inhibition of metabolic cooperation by unsaturated fatty acids.

The effects of certain prostaglandins themselves on metabolic cooperation were also examined (Table 4). Two prostaglandins, 16, 16-dimethylprostaglandin E_2 (PGE_2)

Table 2. Recovery of $6TG^r$ cells in the presence of TPA, and archidonic acid.[a]

Compound	Concentration (μg/ml)	Recovery of $6TG^r$ cells	% Revovery[b]	% Survival[c]
--	--	12.4 ± 1.0	12.8 ± 1.1	--
TPA	0.002	64.6 ± 2.4[d]	66.6 ± 2.5[d]	--
Arachidonic Acid	12.3	28.3 ± 1.5[d]	29.2 ± 1.5[d]	89.2 ± 10.3

[a]. Aylsworth et al (see ref. 137)
[b]. Metabolic cooperation: % Recovery of 100 $6TG^r$ ($HG-PRT^-$) cells co-cultured with 4×10^5 V79 ($HG-PRT^+$) cells relative to the recovery of untreated of untreated $6TG^r$ ($HG-PRT^-$) cells cultivated in the absence of V79 ($HG-PRT^+$) cells. (i.e., plating efficiency)
[c]. Cytotoxicity % Recovery of $6TG^r$ ($HG-PRT^-$) cells relative to the recovery of untreated $6TG^r$ ($HG-PRT^-$) cells cultured alone (i.e., plating efficiency)
[d]. $p < 0.01$ vs. solvent control.

and prostaglandin-A_1-glutathione adduct (PGA_1-glutathione adduct) were tested for their ability to block metabolic cooperation. PGE_2 was unable to inhibit metabolic cooperation in these Chinese hamster V79 cells under the conditions of the assay. On the other hand, PGA_1-11-gluta-thione adduct inhibited metabolic cooperation. It appears that the inhibition of metabolic cooperation occurred at cytotoxic doses. However, one has to remember that in determining the cytotoxicity, only 100 $6TG^r$ cells were used and in the metabolic cooperation plates, 4×10^5 $6TG^s$ cells were present. Thus, the cytotoxicity does not truly reflect the cytotoxicity of PGA_1-11-glutathione adduct in the cocultivation plates. Moreover, although the concentrations used appear extremely high, localized concentrations in tissues might be in that range.

Table 3. Recovery of 6TG^r cells in the presence of TPA, and palmitoleic acid.[a]

Compound	Concentration (µg/ml)	Recovery of 6TG^r cells	% Recovery[b]	% Survival[c]
--	--	12.9 ± 1.3	15.2 ± 1.5	--
TPA	0.002	60.8 ± 2.2[d]	71.9 ± 2.6[d]	--
Palmitoleic Acid	7.5	25.2 ± 1.1[d]	29.8 ± 1.3[d]	86.2 ± 6.9

[a]. Aylsworth et al (see ref. 139)
[b]. Metabolic cooperation: % Recovery of 100 6TG^r (HG-PRT$^-$) cells co-cultured with 4 x 10^5 V79 (HG-PRT$^+$) cells relative to the recovery of untreated 6TG^r (HG-PRT$^-$) cells cultured in the absence of V79 (HG-PRT$^+$) cells (i.e., plating efficiency)
[c]. Cytotoxicity: % Recovery of 6TG^r (HG-PRT$^-$) relative to the recovery of untreated 6TG^r (HG-PRT$^+$) cells cultured alone (i.e., plating efficiency)
[d]. $p < 0.01$ vs. solvent control.

Since there are numerous forms of prostaglandins, and each of them induces different physiological changes in various tissues, more prostaglandins will have to be tested in different cell types to find out if prostaglandins might be physiologically involved in inhibition of intercellular communication.

As stated above, many tumor promoters, including TPA and teleocidin, elevate prostaglandin biosynthesis in many *in vitro* and *in vivo* systems. These compounds are also very effective inhibitors of metabolic cooperation (18,40). Therefore, prostaglandins may also have a role in mediating the inhibition of metabolic cooperation by these and other tumor promoters. We are currently examining the effects of these compounds on prostaglandin synthesis in V79 cells. Furthermore, we also are investigating the possible role of prostaglandins in the inhibition of metabolic cooperation by TPA, teleocidin and the

Table 4. Recovery of $6TG^r$ cells in the presence of TPA and prostaglandins[1].

Compound	Concentration (μg/ml)	% Recovery[a]	% Survival[b]
--	--	18.5 ± 1.02	--
TPA	0.001	86.5 ± 4.9	--
PGE_2	1	18.9 ± 2.5	81.5 ± 3.1
	3	20.7 ± 2.4	68.4 ± 4.0
	5	12.0 ± 1.4	15.4 ± 3.8
PGA_1-Glutathione Adduct	5	27.4 ± 2.8	77.0 ± 1.9
	8	35.3 ± 2.8	67.8 ± 4.7
	10	38.9 ± 3.1	50.9 ± 3.2

[a] = Metabolic cooperation: % Recovery of 100 $6TG^r$ ($HG-PRT^-$) cells cocultivated with 4×10^5 V79 ($HG-PRT^+$) cells relative to the recovery of untreated $6TG^r$ ($HG-PRT^-$) cells cultivated in the absence of V79 ($HG-PRT^+$) cells [i.e., plating efficiency].

[b] = Cytotoxicity: % Recovery of $6TG^r$ ($HG-PRT^-$) cells relative to the recovery of untreated $6TG^r$ ($HG-PRT^-$) cells cultivated alone (i.e., plating efficiency).

[1]. Prostaglandins were generously supplied to us by Dr. Mary Ruwart, the UpJohn Company, Kalamazoo, Michigan.

unsaturated fatty acid precursors utilizing various inhibitors of prostaglandin synthesis. Preliminary data indicate that the prostaglandin synthesis inhibitor indomethacin has no influence on the inhibition of metabolic cooperation by linoleic acid (data not shown).

References

1. Brody H: A systems view of man: Implications for medicine, science and ethics. Perspect Biol Med (17):71-92, 1973.
2. Potter V R: Probabilistic aspects of the human cybernetic machine. Perspect Biol Med (17):164-183, 1974.
3. Loewenstein W R: Junctional intercellular communication and the control of growth. Biochim Biophys Acta (560)
4. Potter V R: Initiation and promotion in cancer formation: The importance of studies on intercellular communication. Yale J Biol Med (53):367-384, 1980.
5. Sachs L: Constitutive uncoupling of pathways of gene expression that control growth and differentiation in myeloid leukemia: A model for the origin and progression of malignancy. Proc Natl Acad Sci USA (77): 6152-6156, 1980.
6. Potter V R: Cancer as a proplem in intercellular communication: Regulation in growth-inhibiting factors (chalones). In: Cohn W E (ed) Progress in Nucleic Acid Research and Molecular Biology. Academic Press, New York, in press.
7. Loewenstein W R: Introductory remarks to the symposium. In Vitro (16):1007-1009, 1980.
8. Rose B: Permeability of the cell to cell membrane channel and its regulation in an insect cell junction. In Vitro (16):1029-1042, 1980.
9. Flagg-Newton J L: The permeability of the cell to cell membrane channel and its regulation in mammalian cell junctions. In Vitro (16):1043-1048, 1980.
10. Pitts J D: The role of junctional communication in animal tissues. In Vitro (16):1049-1056, 1980.
11. Iverson O H: Cybernetic aspects of the cancer problem. In: Weiner N, Schade, J P (eds) Progress in Biocybernetics, Vol. 2. Elsevier Publ. Comp., Amsterdam, 1965, pp 76-110.
12. Trosko J E, Chang C C, Medcalf A: Mechanisms of tumor promotion: Potential role of intercellular communication. Cancer Invest, in press.

13. Trosko J E, Jone C, Chang C C: The role of tumor promoters on phenotypic alterations affecting intercellular communication and tumorigenesis. Ann New York Acad Sci (407):316-327, 1983.
14. Pitot H C, Goldsworthy T, Moran S: The natural history of carcinogenesis: Implications of experimental carcinogenesis in the genesis of human cancer. J Supramol Struct Cell Biochem (17):133-146, 1981.
15. Trosko J E, Chang C C: The integrative hypothesis linking cancer, diabetes and atherosclerosis: The role of mutations and epigenetic changes. Med Hypoth (6): 455-468, 1980.
16. Potter V R: A new protocol and its rationale in the study of initiation and promotion of carcinogenesis in rat liver. Carcinogenesis 2:1375-1379, 1981.
17. Hennings H, Shores R, Wenk L M, Spangler E F, Tarone R, Yuspa S H: Malignant conversion of mouse skin tumours is increased by tumor initiators and unaffected by tumor promoters. Nature (304): 67-69, 1983.
18. Yotti L P, Chang C C, Trosko J E: Elimination of metabolic cooperation in Chinese hamster cells by a tumor promoter. Science (206):1089-1091, 1979.
19. Murray A W, Fitzgerald D J: Tumor promoters inhibit metabolic cooperation in cocultures of epidermal and 3T3 cells. Biochem Biophys Res Commun (91):395-401, 1979.
20. Fitzgerald D J, Murray A N: Inhibition of intercellular communication by tumor-promoting phorbol esters. Cancer Res (40):2935-2937, 1979.
21. Umeda M, Noda K, Ono T: Inhibition of metabolic cooperation in Chinese hamster cells by various chemicals including tumor promoters. Gann (71):614-620, 1980.
22. Trosko J E, Dawson B, Yotti L P, Chang C C: Saccharin may act as a tumor promoter by inhibiting metabolic cooperation between cells. Nature (284) 109-110, 1980.

23. Guy G R, Tapley P M, Murray A W: Tumor promoter inhibition of intercellular communication between cultured mammalian cells. Carcinogenesis (2):223-227, 1981.
24. Enomoto T, Susaki Y, Shiba Y, Kanno Y, Yamasaki H: Tumor promoters cause a rapid and reversible inhibition of the formation and maintenance of electrical cell coupling in culture. Proc Natl Acad Sci USA (78):5628-5632, 1981.
25. Enomoto T, Sasaki Y, Shiba Y, Kanno Y, Yamasaki H: Inhibition of the formation of electrical cell coupling of FL cells by tumor promoters. Gann (72):631-634, 1981.
26. Newbold R F, Amos J: Inhibition of metabolic cooperation between mammalian cells in culture by tumor promoters. Carcinogenesis (2):243-249, 1981.
27. Williams G M, Telang S, Tong C: Inhibition of intercellular communication between liver cells by the liver tumor promoter 1,1,1-tri-chloro 2,2-bis (p-chlorophenyl) ethane. Cancer Letters (11):339-344, 1981.
28. Kinsella A R: Investigation of the effects of the phorbol ester TPA on carcinogen-induced forward mutagenesis to 6-thioguanine-resistance in V79 Chinese hamster cells. Carcinogenesis (2):43-47, 1981.
29. Trosko J E, Yotti L P, Warren S T, Tsushimoto G, and Chang C C: Inhibition of cell-cell communication by tumor promoters. In: Hecker E, Fusenig N E, Kunz W, Marks F, Thielmann H W (eds) Carcinogenesis, Vol. 7. Raven Press, NY. 565-585, 1981.
30. Noda K, Umeda M, Ono T: Effects of various chemicals including bile acids and chemical carcinogens on the inhibition of metabolic cooperation. Gann (72):772-776, 1981.
31. Trosko J E, Dawson B, Chang C C: PBB inhibits metabolic cooperation in Chinese hamster cells in vitro: Its potential as a tumor promoter. Environ Health Perspect (37):179-182, 1981.

32. Slaga T J, Klein-Szanto, ASP, Triplett L C, Yotti, L P, Trosko J E: Skin tumor promoting activity of benzoyl peroxide: A widely used free radical generating compound. Science (213):1023-1025, 1981.
33. Tsushimoto G, Chang C C, Trosko J E, Matsumura F: Cytotoxic, mutagenic and tumor-promoting properties of DDT, Lindane and Chlordane on Chinese hamster cells in vitro. Arch Environ Contam Toxicol, in press.
34. Fitzgerald D J, Murray A W: A new intercellular communication assay: Its use in studies on the mechanism of tumor promotion. Cell Biol Int Rep (6):235-242, 1982.
35. Malcolm A R Mills L d, McKenna E J: Inhibition of metabolic cooperation between Chinese hamster V79 cells by tumor promoters and other chemicals. Ann. New York Acad Sci (407):445-450, 1983.
36. Mosser D D, Bols N C: The effect of phorbols on metabolic cooperation between human fibroblasts. Carcinogenesis (3):1207-1212, 1982.
37. Tsushimoto G, Trosko J E, Chang C C, and Matsumura F: Inhibition of intercellular communication by chlordecone (Kepone) and Mirex in Chinese hamster V79 cells in vitro. Toxicol Appl. Pharmacol (64):550-556, 1982.
38. Tsushimoto G, Trosko J E, Chang C C, Aust S: Inhibition of metabolic cooperation in Chinese hamster V79 cells in culture by various polybrominated biphenyl (PBB) congeners. Carcinogenesis (3) 181-185, 1982.
39. Warren S T, Doolittle D J, Chang C C, Goodman J I, Trosko J E: Evaluation of the carcinogenic potential of 2,4-dinitrofluorobenzene and its implications regarding mutagenicity testing. Carcinogenesis (3):139-145, 1982.
40. Jone C M, Trosko J E, Chang C C, Fujiki H, Sugimura T: Inhibition of intercellular communication in Chinese hamster V79 cells by teleocidin. Gann (73):874-878, 1982.
41. Tsushimoto G, Asano S, Trosko J E, Chang C C: Inhibition of intercellular communication by various congeners of polybrominated biphenyl and polychlorinated biphenyl.

In Dimtri F, Kamrin M (eds) PCB's: Human and Environmental Hazards. Ann Arbor Science Publ., Ann Arbor, 1983, pp. 241-252.

42. Friedman E A, Steinberg M: Disrupted communication between late-stage premalignant human colon epithelial cells by 12-O-tetradecanoyl phorbol-13-acetate. Cancer Res (42):5096-5105, 1982.

43. Yancy S B, Edens J E, Trosko J E, Chang C C, Revel J-P: Decreased incidence of gap junctions between Chinese hamster V79 cells upon exposure to the tumor promoter 12-O-tetradecanoyl phorbol-13-acetate. Exp Cell Res (139):329-340, 1982.

44. Borek C, Sachs L: The difference in contact inhibition of cell replication between normal cells and cells transformed by different carcinogens. Proc Natl Acad Sci USA (56):1705-1711, 1966.

45. Corsaro C M, Migeon B R: Comparison of contact-mediated communication in normal and transformed human cells in culture. Proc Natl Acad Sci USA (74):4476-4480, 1977.

46. Trosko J E, Chang, C C: Error-prone DNA repair and replication in relation to malignant transformation. Transplant Proc., in press.

47. Trosko J E, Chang C C: Role of intercellular communication in tumor promotion. In: Slaga T J (ed). Tumor Promotion and Cocarcinogenesis *in vitro*. CRC Press, Inc. Boca Raton, in press.

48. Trosko J E, Chang C C: A possible mechanistic link between teratogenesis and carcinogenesis: Inhibited intercellular communication. In: Chu EHY (ed). Principles of Mutagenesis, Carcinogenesis and Terestogenesis. Plenum Press, New York, in press.

49. Cooper G M: Cellular transforming genes. Science (218): 801-806, 1982.

50. Bishop J M: Oncogenes. Scientific American (246):81-92, 1982.

51. Hausen Z H, O'Neill F J, Freese U K, Hecker E: Persisting oncogenic herpes virus induced by the tumor promoter TPA. Nature (272):373-375, 1978.

52. Yamamoto N, Hausen Z H: Tumor promoter TPA enhances transformation of human leukocytes by Epstein-Barr virus. Nature (280):244-245, 1979.
53. Colletta G, DiFiore P P, Ferrentino M, Pietropaolo C, Turco M C, Vecchio G: Enhancement of viral gene expression in Friend erthroleukemic cells by 12-0-tetradecanoyl phorbol-13-acetate. Cancer Res (40): 3369-3373, 1980.
54. Arya S K: Phorbol ester-mediated stimulation of synthesis of mouse mammary tumour virus. Nature (284): 71-72, 1980.
55. Hellman K B, Hellman A: Induction of type-C retrovirus by the tumor promoter TPA. Int J Cancer (27):95-99, 1981.
56. Yamamoto H, Katsuki T, Hinuma Y, Hoshino H-o, Miwa M, Fujiki H, Sugimura T: Induction of Epstein-Barr virus by a new tumor promoter teleocidin, compared to induction by TPA. Int J Cancer (28):125-129, 1981.
57. Amtmann E, Sauer G: Activation of non-expressed bovine papilloma virus genomes by tumor promoters. Nature (296):675-676, 1982.
58. Brugge J S, Erikson R L: Identification of a transformation-specific antigen induced by an avian sarcoma virus. Nature (269):346-348, 1977.
59. Levinson A D, Oppermann H, Levinton L, Varmus H E, Bishop J M: Evidence that the transforming gene of avian sarcoma virus encodes a protein kinase associated with a phosphoprotein. Cell (15):561-572, 1978.
60. Erikson R L, Collett M S, Erikson E, Purchio A F: Evidence that the avian sarcoma virus transforming gene product is a cyclic AMP-independent protein-kinase. Proc Natl Acad Sci USA (76):6260-6224, 1979.
61. Epstein J, Breslow J L, Fontaine J H: Enhanced phosphorylation of many endogenous protein substrates in human fibroblasts transformed by simian virus 40. Proc. Natl Acad Sci USA (76):6396-6400, 1979.

62. Laszlo A, Radke K, Chin S, Bissell M J: Tumor promoters alter gene expression and protein phosphorylation in avian cells in culture. Proc Natl Acad Sci USA (78):6241-6245, 1981.
63. Cooper J A, Reiss N A, Schwartz R J, Hunter J: Three glycolytic enzymes are phosphorylated at tyrosine in cells transformed by Rous sarcoma virus. Nature (302): 218-223, 1983.
64. Kraft A S, Anderson W B: Phorbol ester increase the amount of Ca^{2+}, phospholipid-dependent protein kinase associated with plasma membrane. Nature (301):621-623, 1983.
65. Niedel J E, Kuhn L J, Vandenburk G R: Phorbol diester receptor copurifies with protein kinase C. Proc Natl Acad Sci USA (80):36-40, 1983.
66. Atkinson M M, Mouko A S, Johnson R G, Sheppard J R, Sheridan J D: Rapid and reversible reduction of junctional permeability in cells infected with a temperature-sensitive mutant of avian sarcomas virus. J Coll Biol (91):573-578, 1981.
67. Karim S M M, Rao B. Prostaglandins and tumors. In: Karim S M M (ed.) Advances in Prostaglandin Research. Prostaglandins: Physiological, Pharmacological and Pathological aspects. MTP Press, London, 1976, pp. 301-325.
68. Levine L. Effects of tumor promoters on arachidonic acid metabolism by cells in culture. In: Hecker E, Fusenig N E, Kunz W, Marks F, Thielmann H W (eds.) Carcinogenesis Vol. 7. Cocarcinogenesis and Biological Effects of Tumor Promoters. Raven Press, New York, 1982, pp. 477-494.
69. Form D M, Sidky Y A Kubai L, Auerback R. PGE_2-induced angiogenesis. In: Powles T J, Bockman R S, Honn K V, Ramwell P (eds.) Prostaglandins and Cancer: First International Conference, Alan R. Liss Inc., New York, 1982, p. 685.

70. Rolland P H, Martin P M, Jacquemier J, Rolland A M, Toga M: Prostaglandin in human breast cancer: Evidence suggesting that an elevated prostaglandin production is a marker of high metastatic potential for neoplastic cells. J Natl Cancer Inst (64):1061-1070, 1980.
71. Form D M, Auerbach R: PGE_2 and Angiogenesis. Proc Soc Expt Biol Med (172):214-218, 1983.
72. Bennett A, McDonald A M, Stamford I F, Charlier E M, Simpson J S, Zbro T: Prostaglandins and breast cancer. Lancet (1):624-626, 1977.
73. Bennett A, Simpson J S, McDonald A M, Stamford I F: Breast cancer, prostaglandins, bone metastasis. Lancet (2):1218-1220, 1975.
74. Honn K V, Bockman R S, Marnett L J: Prostaglandins and cancer: A review of tumor initiation through tumor metastasis. Prostaglandins (21):833-864, 1981.
75. Jaffe B M: Prostaglandins and cancer: An update. Prostaglandins (6):453-461, 1974.
76. Bennett A, Carroll M A, Stamford I F, Whimster W F, Williams F: Prostaglandins and human lung carcinomas. Br J Cancer (46):888-893, 1982.
77. Humes J L, Strausser H R: Prostaglandins and cyclic nucleotides in maloney sarcoma tumors. Prostaglandins (5):183-196, 1974.
78. Tashjian A H, Voelkel E F, Goldhaber P, Levine L: Successful treatment of hypercalcemia by indomethacin in mice bearing a prostaglandin-producing fibrosarcoma. Prostaglandins (3):315-524, 1973.
79. Sykes J A C, Maddox J: Prostaglandin production by experimental tumors and effects of anti-inflammatory compounds. Nature (London), New Biol (237):59-61, 1972.
80. Tan W C, Privett O S, Goldyne M E: Studies of prostaglandins in rat mammary tumors induced by 7,12-dimethylbenz(a)anthracene. Cancer Res (34):3229-3231, 1974.
81. Tashjian A H, Voekel E F, Levine L: Plasma concentrations of 13,14-dihydro-15-keto-prostaglandin E_2 in

rabbits bearing VX_2 carcinoma: Effects of hydrocortisone and indomethacin. Prostaglandins (14):309-317, 1977.

82. Levine L, Hinkle P M, Voelkel E F, Tashijian A H: Prostaglandin production by mouse fibrosarcoma cells in culture: Inhibition by indomethacin and aspirin. Biochem Biophys Res Commun (47):888-891, 1972.

83. Ashendel C L, Boutwell R K: Prostaglandin E and F levels in mouse epidermis are increased by tumor-promoting phorbol esters. Biochem Biophys Res Commun (90):623-627, 1979.

84. Bresneck E, Meunier P, Lamden M: Epidermal prostaglandins after topical application of a tumor promoter. Cancer Lett. (7):121-125, 1979.

85. Tashjian A H, Ivey J L, Delclos B, Levine L: Stimulation of prostaglandin production in bone by phorbol esters and melittin. Prostaglandins (16):221-232, 1978.

86. Crutchley D J, Conanan L B, Maynard J R: Induction of plasminogen activator and prostaglandin biosynthesis in Hela cells by 12-0-tetradecanoyl phorbol-13-acetate. Cancer Res (40):849-852, 1980.

87. Hammarström S, Lindgren J A, Marcelo C, Duell E A, Anderson ThF, Voorhees J J: Arachidonic acid transformation in normal and psoriatic skin. J Invest Dermatol (73):180-183, 1979.

88. Fürstenberger G, Richter H, Fuseniq N E, Marks F: Arachidonic acid and prostaglandin E_2 release and enhanced cell proliferation induced by the phorbol ester TPA in a murine epidermal cell line. Cancer Lett. (11): 191-198, 1981.

89. Levine L: Stimulation of cellular prostaglandin production by phorbol-esters and growth factors and inhibition by chemopreventive agents. In: Powles T J, Bockman R S, Honn R V, Ramwell P (eds.). Prostaglandins and Cancer: First International Conference. Alan R. Liss Inc., New York. pp. 189-204, 1982.

90. Verma A K, Ashendel C L, Boutwell R K: Inhibition by prostaglandin synthesis inhibitors of the induction of epidermal ornithine decarboxylase activity, the accumulation of prostaglandins and tumor promotion caused by 12-0-tetradecanoylphorbol-13-acetate. Cancer Res (40):308-315, 1980.

91. Verma A K, Rice H M, Boutwell R K: Prostaglandins and skin tumor promotion: Inhibition of tumor promoter-induced ornithine decarboxylase activity in epidermis by inhibitors of prostaglandin synthesis. Biochem Biophys Res Commun (79):1160-1166, 1977.

92. Ohuchi K, Levine L: Tumor promoting phorbol esters stimulate release of radioactivity from [^{3}H]-arachidonic acid labeled - but not [^{14}C] linoleic acid labeled - cells. Indomethacin inhibits the stimulated release from [^{3}H] arachidonate labeled cells. Prostaglandins and Medicine 1:421-431, 1978.

93. Sakamoto H, Terada M, Fujiki H, Mori M, Nakayasu M, Sugimura T, Weinstein L B: Stimulation of prostaglandin production and choline turnover in HeLa cells by lyngbyatoxin A and dihydroteleocidin B. Biochem Biophys Res Commun (102):100-107, 1981.

94. DeAsua L J, Clingan D, Rudland P S: Initiation of cell proliferation in cultured mouse fibroblasts by prostaglandin F2α. Proc Natl Acad Sci (72):2724-2728, 1975.

95. Karmali R A, Horrobin D F, Manezes J, Patel P: The relationship between concentrations of prostaglandins A,E,E_2 and F2α and rates of cell proliferation. Pharm Res Comm (11):69-75, 1979.

96. Stobo J D, Kennedy M S, Goldyne M E: Prostaglandin E modulation of the mitogenic response of human T-cells. J Clin Invest (64):1188-1195, 1979.

97. Saez J M, Evain D, Gallet D: Role of cyclic AMP and protein kinase on the steroidogenic action of ACTH, prostaglandin E_1 and dibutyryl cyclic AMP in normal

adrenal cells and adrenal tumor cells from humans. J Cyclic Nucleotide Res (4):311-321, 1978.

98. Santoro M G, Philpott G W, Jaffe B M: Inhibition of tumor growth *in vivo* and *in vitro* by prostaglandin E. Nature (263):777-779, 1976.

99. Smith J W, Steiner A L, Parker C W: Human lymphocyte metabolism: Effects of cyclic and noncyclic nucleotides on stimulation by phytohemagglutinin. J Clin Invest (50):442-448, 1971.

100. Thomas D R, Philpott G W, Jaffe B M: The relationship between concentration of prostaglandin E and rates of cell replication. Exptl Cell Res (84):40-46, 1974.

101. Claesson H-E, Lindgren J A, Hammarstrom S: Endogenous prostaglandin E_2 production inhibits proliferation of polyoma virus-transformed 3T3 cells: Correlation with cellular levels of cyclic AMP. Adv Prost Thromb Res (6):541-546, 1980.

102. Fürstenberger G, Marks F: Indomethacin inhibition of cell proliferation induced by the phorbol ester TPA is reversed by prostaglandin E_2 in mouse epidemis *in vivo*. Biochem Biophys Res. Comm (84):1103-1111, 1978.

103. Fürstenberger G, Marks F: Early prostaglandin E synthesis is an obligatory event in the induction of cell proliferation in mouse epidermis *in vivo* by the phorbol ester TPA. Biochem Biophys Res Commun (92):749-756, 1980.

104. Hammarström S, Hamberg M, Duell E A, Stawiski M A, Anderson T F, Voorhees J J: Glucocorticoid in inflammatory proliferative skin disease reduces arachidonic and hydroxyeicosatetraenoic acids. Science (197):994-996, 1977.

105. Hammarström S, Hamberg M, Samuelsson B, Duell E A, Strawiski M, Voorhees J J: Increased concentrations of nonesterifed arachidonic acid, 12L-hydroxy-5,8,10,14-eicosatetraenoic acid, prostaglandin E_2 and prostaglandin Fα in epidermis of psoriasis. Proc Natl Acad Sci (72):5130-5134, 1975.

106. MacManus J P, Braceland B W: A connection between the production of prostaglandins during liver regeneration and the DNA synthetic response. Prostaglandins (11): 609-620, 1976.
107. Tutton P J M, Barkla D H: Influence of prostaglandin analogues on epithelial cell proliferation and xenograft growth. Br J Cancer (41):47-51, 1980.
108. Tashjian A M, Voelkel E F, Goldhaber P, Levine L: Prostaglandins, calcium metabolism and cancer. Fed. Proc. (33):81-86, 1974.
109. Humes J L, Cupo J J, Strausser H R: Effects of indomethacin on Moloney sarcoma virus-induced tumors. Prostaglandins (6):463-473, 1974.
110. Lynch N R, Salomon J C: Tumor growth inhibition and potentiation of immuno-therapy by indomethacin in mice. J Natl Cancer Inst (62):117-121, 1979.
111. Trevisani A, Ferretti E, Capuzzo A, Tomasi V: Elevated levels of prostaglandin E_2 in Yoshida hepatoma and the inhibition of tumour growth by non-steroidal anti-inflammatory drugs. Br J Cancer (41):341-347, 1980.
112. Hial V, Horakova Z, Shaff R E, Beaven M A: Alteration of tumor growth by aspirin and indomethacin: Studies with two transplantable tumors in mice. Eur J Pharmacol (37):367-376, 1976.
113. Leaper D J, French B T, Bennett A: Breast cancer and prostaglandins: A new approach to treatment. Br J Surg (66):683-686, 1979.
114. Strausser H R, Humes J L: Prostaglandin synthesis inhibition: Effects on bone changes and sarcoma tumor induction in BALB/c mice. Int J Cancer (15):724-730, 1975.
115. Bennett A, Houghton J, Leaper D J, Stamford I F: Cancer growth, response to treatment and survival time in mice: Beneficial effect of the prostaglandin synthesis inhibitor fluribiprofen. Prostaglandins (17):179-191, 1979.

116. Lynch N R, Castes M, Astoin M, Salomon J C: Mechanism of inhibition of tumor growth by aspirin and indomethacin. Br J Cancer (38):503-512, 1978.
117. Tobias L D, Hamilton J G: The effect of 5,8,11,14-Eicosatetraynoic acid on lipid metabolism. Lipids (14):181-193, 1979.
118. Plecia O J, Smith A H, Grinwich K: Subversion of immune system by tumor cells and role of prostaglandins. Proc Natl Acad Sci (72):1848-1851, 1975.
119. Hillyard L A, Abraham S: Effect of dietary polyunsaturated fatty acids on growth of mammary adenocarcinomas in mice and rats. Cancer Res (39):4430-4437, 1979.
120. Rao G A, Abraham S: Brief communication: Reduced growth rate of transplantable mammary adenocarcinoma in C_3H mice fed eicosa-5,8,11,14-tetraynoic acid. J Natl Cancer Inst (58):445-447, 1977.
121. Bennett A, Houghton J, Leaper D J, Stamford I F: Tumour growth and response to treatment: Beneficial effect of the prostaglandin synthesis inhibitor fluribiprofen. Br J Pharmac (63):356P-357P, 1978.
122. Bennett A, Berstock D A, Carroll M A: Increased survival of cancer-bearing mice treated with inhibitors of prostaglandin synthesis alone or with chemotherapy. Br J Cancer (45):762-768, 1982.
123. Lupulescu A: Effects of prostaglandins on tumor transplantation. Oncology (37):418-423, 1980.
124. Jubiz W, Frarley J, Smith J B: Inhibitory effect of prostaglandin $F_{2\alpha}$ on the growth of a hormone-dependent rat mammary tumor. Cancer Res (39):998-1000, 1979.
125. Berenblum I: A re-evaluation of the concept of cocarcinogenesis. Prog Exp Tumor Res (11):21-30, 1969.
126. Viaje A, Slaga T J, Wigler M, Weinstein, I B: Effects of anti-inflammatory agents on mouse skin tumor promotion, epidermal DNA synthesis, phorbol ester-induced cellular proliferation, and production of plasminogen activator. Cancer Res (37):1530-1536, 1977.

127. Slaga T J, Scribner J D: Inhibition of tumor initiation and promotion by anti-inflammatory agents. J Natl Cancer Inst (51):1723-1725, 1975.
128. Slaga T J, Fischer S M, Viaje A, Berry D L, Bracken W M, LeClerc S, Miller D R: Inhibition of tumor promotion by anti-inflammatory agents: An approach to the biochemical mechanism of promotion. In: Slaga T J, Sivak A, Boutwell R K (eds) Carcinogenesis Vol 2: Mechanisms of Tumor Promotion and Cocarcinogenesis. Raven Press, New York, 1978, pp. 173-175.
129. Fischer S M, Mills G D, Slaga T J: Inhibition of mouse skin tumor promotion by prostaglandin and thromboxane synthesis inhibitors. Carcinogenesis (3):1243-1245, 1982.
130. Fischer S M, Gleason G L, Mills G D, Slaga T J: Indomethacin enhancement of TPA tumor promotion in mice. Cancer Lett (10):343-350, 1980.
131. Fischer S M, Gleason G L, Hardin L G, Bohrman S, Slaga T J: Prostaglandin modulation of phorbol ester skin tumor promotion. Carcinogenesis (1):245-248, 1980.
132. Kudo T, Narisawa T, Abo S. Antitumor activity of indomethacin on methyl-azoxymethanol-induced large bowel tumors in rats. Gann (71):260-264, 1980.
133. Lupulescu A: Enhancement of carcinogenesis by prostaglandins. Nature (272):634-636, 1978.
134. Troll W, Belman S, Goldstein, B, Mukai F, Machlin L: Effect of feeding unsaturated or saturated fat on carcinogenesis on mouse skin. Proc Am Assoc Cancer Res (19):106, 1978.
135. Carter C A, Milholland R J, Shea W, Ip M M: Effect of the prostaglandin synthesis inhibitor indomethacin on 7,12-dimethylbenz(a)anthracene-induced mammary tumorigenesis in rats fed different levels of fat. Cancer Res (43):3559-3562, 1983.
136. Lupulescu A: Enhancement of carcinogenesis by prostaglandins in male albino Swiss mice. J Natl Cancer Inst (61):97-106, 1978.

137. Brune K, Kalin H: Inflammatory, tumor-initiating and promoting activities of polycyclic aromatic hydrocarbons and diterpene esters in mouse skin as compared with their prostaglandin releasing potency in vitro. Cancer Lett (4):333-342, 1978.
138. Trosko J E, Yotti L P, Dawson B, Chang C C: In vitro assay for tumor promoters. In: Stich H, San R H C (eds) In Vitro Tests for Chemical Carcinogens. Springer-Verlag, New York, 1981, pp. 420-427.
139. Aylsworth C F, Jone C, Trosko J E, Welsch C W: Influence of dietary fatty acids on metabolic cooperation by Chinese hamster V79 cells in vitro. (in preparation)
140. Trosko J E, Jone C, Aylsworth C F, Tsushimoto G: Elimination of metabolic cooperation is associated with tumor promoters oleic acid and anthralin. Carcinogenesis (3):1101-1103, 1982.

Chapter 8

PROTEASES AND CYCLIC NUCLEOTIDES
Sidney Belman and Seymour Garte
New York University Medical Center
New York, N.Y. 10016

Page

I. Proteases

  A. Introduction 200

  B. Use of proteolytic inhibitors
    1. Effects on tumor induction in vivo. 200
    2. Effects of protease inhibitors in vitro 206
    3. Effects of protease inhititors on growth of
       tumor cells 210
    4. Effects of protease inhibitors on properties of
       transformed and untransformed cells in vitro. 211
    5. The role of proteases in the association of
       superoxide and peroxides with promotion 213
    6. Role of protease. .219
    7. Role of proteases in promotion and carcinogenesis 221

  C. Summary. 227

II. Cyclic Nucleotides. .227

References. 233

S.M. Fischer and T.J. Slaga (eds.), ARACHIDONIC ACID METABOLISM AND TUMOR PROMOTION. Copyright © 1985. Martinus Nijhoff Publishing, Boston. All rights reserved.

8. PROTEASES AND CYCLIC NUCLEOTIDES

S. BELMAN AND S.J. GARTE

I. PROTEASES

A. INTRODUCTION

Recent symposia and reviews (1-8) have attributed to proteases a significant role in regulation of physiological functions. Aside from the well-known function of general protein degradation, they also produce limited proteolysis which converts zymogens to active enzymes by the hydrolysis of a single peptide bond or of several bonds by sequential action. Much work over the past 10-15 years has been directed at the determination of a specific role for proteases in the regulation of normal cell growth, the functions of specialized cells, and in the control of tumor development, growth and metastases. The presence of a variety of protease inhibitors with varying types of specificity in plasma and cells of many tissues (8) points to a fine control of physiological functions for proteases.

This review focuses on evidence that assigns an important role to proteases in promotion and carcinogenesis. Some references to particular articles are given in reviews, symposia, and books, which should be consulted by the reader.

B. USE OF PROTEOLYTIC INHIBITORS

1. Effects on tumor induction in vivo

The first indication that proteases affect tumor promotion were observations (9,10) that synthetic and natural protease inhibitors counteracted the effect of the model promoter, phorbol-myristate-acetate (PMA) or croton oil, on mouse skin. This has led to a search for a possible mechanism for proteases in tumorigenesis.

Diverse types of protease inhibitors are being used to find a function of proteases in tumorigenesis. These include the polypeptides, soybean trypsin and bovine pancreatic trypsin inhibitors, the synthetic chloromethyl ketones, and the small microbial polypeptide aldehydes.

The effects of protease inhibitors on tumor induction in vivo are summarized in Table 1. Troll and his collaborators (9) were the first to show that protease inhibitors inhibited tumor induction in vivo although others had previously observed inhibition of tumor growth by aprotinin. They observed that the promotion of 7,12-dimethylbenz[a]anthracene (DMBA)-initiated mouse skin tumorigenesis was inhibited by low doses (1-10 µg) of tosyllysine chloromethylketone (TLCK), tosylphenylalanyl chloromethylketone (TPCK) and tosylarginine methyl ester (TAME). The first two agents inhibit trypsin and chymotrypsin, respectively, in addition to papain, and TAME inhibits trypsin and papain. The number and incidence of tumors were decreased and the latent period was increased. The most effective agent was TPCK which almost completely suppressed tumor formation. Ear irritation and leukocyte infiltration by PMA was inhibited by the three protease inhibitors as well (9,11). Local vascular changes, accompanied by mast cell degranulation, which was induced by PMA, was suppressed by TPCK (11). Protease activity, as measured by TAME esterase activity of ear homogenates, was increased 3-5 fold by PMA and was inhibited by 0.1 M TLCK and 0.1 M TPCK. The inhibitory effect of TPCK on promotion by PMA was confirmed, in Ha/ICR mice by Belman (12) and in Skin Tumor Sensitive mice by Slaga et al (13).

Hozumi et al (10) demonstrated that the microbial inhibitor, leupeptin, applied topically, inhibited croton oil-promoted mouse skin tumorigenesis. They observed a decreased incidence and number of tumors as well as an increased latent period. Furthermore, leupeptin inhibited the croton oil-induced mouse skin TAME esterase activity in vitro and in vivo in a dose-dependent manner.

Table 1. Effect of protease inhibitors on tumor induction in vivo.

Protease inhibitor	Effects/carcinogen/cell system[a]	References
TPCK, TAME, TLCK	I/DMBA-croton oil or PMA promotion/mouse skin	(9),(12),(13)
Leupeptin	I/DMBA-croton oil/mouse skin	(10),(14)
Leupeptin	I/azoxymethane/rat colon	(10)
Leupeptin	I/DMBA/rat mammary	(10)
Leupeptin	I/butylnitrosourethan/rat esophagus; NE on forestomach	(10)
Leupeptin	E/MNNG/rat stomach	(10)
Leupeptin	E/X-rays/mice leukemia	(10)
Leupeptin	NE/urethan/mouse lung	(10)
Leupeptin	E and NE N-OH-BBN/rat bladder	(15)
Leupeptin	I/N-OH-BBN/rat bladder	(16)
Leupeptin	NE/DEN/rat liver	(10)
Leupeptin	I/N-nitrosobutylurea/mouse leukemia	(10),(17)
MeDBG	I/DMBA/rat mammary	(18)
EACA	I/DMH/mouse colon	(19)
Leupeptin	E (males only)/spontaneous/mouse liver	(20)
Soy bean diet	I/NQO-PMA/promotion mouse skin	(6)
Soy bean diet	I/X-rays/rat mammary	(2)
Soy bean concentrate	I/spontaneous/male mouse liver	(21)

[a] I=inhibition; E=enhancement; NE=no effect

These observations led to the testing of protease inhibitors in skin and other organs (Table 1). Many of the results are described by Matsushima et al (10). Leupeptin, a potent inhibitor of trypsin, plasmin, and papain, but not of chymotrypsin, had variable effects. It showed inhibition, enhancement, and no effect.

Hosaka and Hirono (20) fed leupeptin to strain A mice which develop spontaneous tumors. There was no effect on lung tumor formation in male and female mice but an effect on liver tumors in male mice was obtained. The leupeptin-fed male mice exhibited a 40% incidence of liver tumors compared with 17.4% in the controls.

In the experiment by Goto et al (14) the initiating dose of 125 µg DMBA was itself tumorigenic since all mice developed many tumors. Croton oil promotion produced about a 2 fold increase in tumor yield. Treatment with 1 mg leupeptin produced a slightly lower tumor incidence and a decreased tumor yield which was approximately that obtained with DMBA alone. There also seemed to be a lag in the time period for increased tumor appearance. This paper shows that leupeptin inhibits the promoting effect of croton oil and suggests that it also suppresses the promoting effect of DMBA.

Yamamura et al (18) tested a synthetic inhibitor, N,N-dimethylamino-[p-(p'-guanidinobenzyloxy)] benzilcarbonyloxy glycolate (MeDBG) which was an effective inhibitor of trypsin-like enzymes. A diet of MeDBG delayed and suppressed rat mammary tumors induced with DMBA.

Corasanti et al (19) obtained inhibition of 1,2-dimethylhydrazine(DMH)-induced mouse colorectal tumors by oral administration of ε-aminocaproic acid (EACA). This protease inhibitor markedly reduced the number of tumors.

The above observations were made with the use of purified low molecular weight protease inhibitors. Three experiments were performed with soy bean diets which contain large polypeptide protease inhibitors. Troll et al (22) fed a casein diet, a roasted and a raw soy bean diet to mice starting at two weeks before initiation with 4-nitroquinoline-N-oxide (NQO). The mice were promoted with PMA. The onset and tumor incidence were not affected by the roasted soy bean diet but the raw soy bean diet produced a delay of about 140 days in onset of skin tumor formation and a decreased tumor incidence. The trypsin inhibitor content of these diets was 2.21, 0.10, and 0.17% for the raw, roasted soy bean, and casein diets, respectively.

Troll and his collaborators (22) also found that a soy bean diet lowers breast tumor incidence in irradiated rats.

A remarkable experiment by Becker (21) demonstrated a dramatic inhibition of spontaneous hepatocarcinogenesis in C3H/HeN mice by a special soy bean diet. He fed a preparation called Edi Pro A, a soy bean-derived protein concentrate which contains 3% of protease inhibitors. Four diets were given which contained various amounts of Edi Pro A. The incidence of spontaneous liver cancer at 18 months was; 100, 25, 12.5, and 0% with diets containing respectively, 0, 2.6, 3.9, and 5.2% Edi Pro A.

Soy bean trypsin inhibitor (SBTI) was also reported (2) to cure mice from ascites tumors. A dose-dependent incidence of cures ranged from 0-40%.

We have been using certain type-specific protease inhibitors in order to gain further insight into the nature of the proteases which may be important in promotion. The results with L-<u>trans</u>-epoxy-succinyl-leucylamido (3-methyl)butane (Ep-475) are shown in Figure 1. This agent is one of a series of related peptides, isolated from <u>Aspergillus japonica</u>, that appears to act as class- specific inhibitors of cysteine proteinases (23). Ep-475 was one of the most active which did not inhibit other proteases, such as trypsin or chymotrypsin. The inhibition of promotion, although small, suggests that cathepsin-like enzymes may be involved in promotion.

Octylisocyanate was shown (24) to selectively inhibit chymotrypsin and papain, while having little or no effect on trypsin, elastase, pepsin, or carboxypeptidase. It appears to be an active-site-specific reagent for chymotrypsin since it was bound to the active-site serine.

Bromoacetone is a halomethylketone which reacts with sulfhydryl and histidine in trypsin which was thereby inactivated (25). A di-halomethylketone, 1,3-dibromo-acetone, rapidly and irreversibly inactivated ficin, stem-bromelain, and papain, which contain active-site cysteines adjacent to histidine. This bifunctional reagent forms cross-links between the 2 amino acids which are in close proximity to each other (26-28).

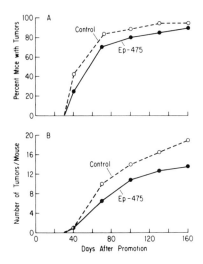

FIGURE 1. Effect of Ep-475. Twenty Ha/ICR mice were in each group. They were initiated with 25 µg DMBA and promoted 3x/week with 5 µg PMA. The control group received acetone while the other was treated with 477 nmol (150 µg) Ep-475, 30 min after PMA treatment.

While these 3 protease inhibitors inhibited promotion (Figure 2), dibromoacetone was much more effective. The dose-dependent inhibition is shown in Figure 2b.

These results suggest that cathepsin-like enzymes play a role in promotion.

There is a distinct difference between large polypeptide protease inhibitors and those of low molecular weight which can readily be absorbed and disseminated throughout the body. Yavelow et al (29) demonstrated that Bowman-Birk soy bean protease inhibitor (BB) is not absorbed by rodent intestine, whereas plasminogen was partially absorbed. The BB was excreted as a complex with digestive proteases while retaining its anti-protease activity. The authors proposed that the anticarcinogenic activity of BB-like polypeptides may result from the consequent inhibition of protein digestion. The direct effect of polypeptides on cells in culture and in situ may hence be different from that obtained by feeding or at distant sites.

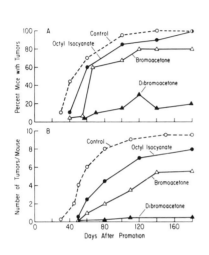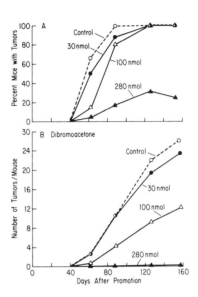

FIGURE 2. Effects of protease inhibitors. The protocol was similar to that in Fig. 1. a) The mice were promoted with 0.2 ml 0.05% croton oil in acetone and the inhibitors were applied 5x/week. The dose per application was; 1.2 µmol octylisocyanate, 1.46 µmol bromoacetone, and 0.93 µmol 1,3-dibromoacetone, b) these mice were promoted with 5 µg PMA.

2. Effects of protease inhibitors in vitro

The transformation of tissue cells in culture to a neoplastic form by chemicals and radiation and by an initiation-promotion protocol, provides a convenient system for examining the tumorigenic process and its alteration by various agents, although it is not a substitute for experimental animals (30).

Several recent reports, summarized in Table 2, have described the effects of protease inhibitors on the transformation of cells in vitro by complete carcinogens and by the initiation-promotion protocol.

Table 2. Effects of protease inhibitors on transformation of cells in vitro.

Protease inhibition/ time added with respect to carcinogen	Effects/carcinogen/ cell system[a]	References
Antipain/24 hrs/BT	E/X-rays/hamster,mouse	(31)
Antipain/10 min BT or 24 hrs AT	I/X-rays/hamster	(31)
Antipain/10 min AT	I/X-rays/mouse	(31)
Antipain/24 hrs or 48 hrs AT	NE/X-rays/mouse,hamster	(31)
Antipain/IM or 4 days AT	I/X-rays/mouse	(32)
Antipain/present 1 or 5 days AT	I/X-rays/mouse	(32)
Leupeptin/IM or for 5 days AT	I/X-rays/mouse	(32)
SBTI/present for 1 day or continuous AT	NE/X-rays/mouse	(32)
Antipain or leupeptin/ IM AT	I/X-rays+PMA/mouse	(32)
Antipain/4 days AT	I/X-rays+PMA/mouse	(32)
SBTI/continuous AT	I/X-rays+PMA/mouse	(32)
Antipain/24 hrs BT, removed 10 min AT	E/X-rays/hamster,mouse	(33)
Antipain/10 min AT, removed 24 hrs AT	I/X-rays/hamster,mouse	(33)
Antipain	I/MNNG/hamster	(34)
Antipain, leupeptin, chymostatin, pepstatin, elastatinal	I/3-MC/mouse	(35)
Antipain, leupeptin	I/17 β-estradiol/mouse	(36)
Antipain, leupeptin	I/X-rays+17 β-estradiol/ mouse	(36)
Chymostatin	I/UV, X-ray, X-ray+PMA/ mouse	(37)

[a] hamster=hamster embryo cells; mouse=mouse C3H/10T½ cells; AT=after treatment; BT=before treatment; IM=immediately. Other abbreviations as in Table 1.

Borek et al (31) and Geard et al (33) found that low concentrations (10 µM) of antipain enhanced or inhibited X-ray-induced transformation in hamster embryo and mouse C3H/10T½ cells. The time of antipain addition with respect to irradiation, and the period of cell contact, were the determining factors. Addition of antipain 24 hrs before irradiation and its continued presence or removal 10 min afterwards resulted in enhancement, whereas addition of antipain 10 min before or after irradiation inhibited transformation. Addition at 24 or 48 hrs after irradiation had no effect.

Antipain, like other microbial inhibitors, is an aldehyde. Its presence before and during X-irradiation may lead to increased active oxygen species which may be responsible for its enhancement of transformation.

Kennedy and Little (32) found that higher concentrations of antipain (0.1 and 1 mM) inhibited transformation of C3H/10T½ cells by X-rays when it was added immediately after or 4 days post-irradiation. They also observed that a 1 or 15 day presence was sufficient to inhibit transformation. They did not add antipain prior to irradiation. Less extensive studies with leupeptin gave results that were similar to those obtained with antipain. These authors also found that antipain markedly suppressed X-ray transformation enhanced by PMA (32,38). There was complete suppression with 400 rads plus PMA. Leupeptin gave variable results (32) but appeared to be less effective in suppressing PMA enhancement of transformation (38). The protein, SBTI, inhibited the PMA-enhanced transformation although it did not affect transformation by X-rays alone.

Kennedy and Little (32) interpreted their data in terms of 3 phases of radiation transformation: a fixation phase which is the 24 hr post-irradiation period when DNA repair processes are operating: an expression phase which extends for 6 weeks following fixation and which is required for the phenotypic manifestations of the transformed state: and a promotion phase within the expression phase that involves enhancement by promoting agents such as PMA. They concluded that the fixation phase was affected by antipain but not by SBTI; the expression phase was affected by antipain and leupeptin but not by SBTI and the promotion phase was affected by antipain (and leupeptin) and SBTI. They surmised from this that various types of proteases may be involved in various phases of transformation. It should be noted that SBTI is probably acting on the cell membrane rather than in the cytoplasm. Geard et al (33) believe that antipain may have several modes of action which include effects on some rapid error-free DNA repair that would be responsible for its enhancement of radiation.

Other studies by Kennedy and Weichselbaum (36) showed that antipain and leupeptin markedly suppressed transformation by β-estradiol and enhanced transformation by β-estradiol and X-rays.

DiPaolo et al (34) found that antipain suppressed N-methyl-N'-nitro-N-nitrosoguanidine (MNNG)-induced transformation in hamster embryo cells in a dose-dependent manner. Kuroki and Devron (35) observed that antipain, chymostatin, elastatinal, leupeptin, and pepstatin were effective inhibitors of 3-methylcholanthrene (3-MC)-induced transformation of mouse C3H/10T½ cells. When removed after 48 hrs only elastatinal appeared to show inhibition. All, except elastatinal, were effective inhibitors when added 1 week after 3-MC treatment. This indicated to the authors that the protease inhibitors were acting at a late promotional stage. It is of interest that chymostatin and pepstatin were most effective. Kuroki and Devron (35) also found that these protease inhibitors did not affect V79 Chinese hamster cell mutagenesis by 3-MC or UV. Only elastatinal appeared to inhibit azaguanine but not ouabain-resistant mutagenesis.

Kennedy (37) found that chymostatin was the most potent inhibitor of transformation by UV, X-rays, and PMA-enhanced X-rays. Chymostatin was also the most effective inhibitor of PMA-stimulated release of plasminogen activator (PA) (39) and superoxide (22).

We may conclude from the in vitro studies that various protease inhibitors can inhibit transformation by chemicals and X-rays and transformation enhanced by promoting agents. The inhibition at later stages of the former also suggest activity at the promotional phase.

3. Effects of protease inhibitors on growth of tumor cells in vivo

The observation by Troll and Hozumi, described above, that protease inhibitors suppressed tumorigenesis in mouse skin, and those of Burger (2) and Sefton and Rubin (1) which demonstrated that proteases induce cell growth in vitro stimulated other workers to test the effects of protease inhibitors on the growth of tumor cells in vivo. Their observations are summarized in Table 3.

Whur et al (40) found that a single or multiple i.p. injections of SBTI inhibited the growth of Ehrlich ascites cells. The increased adhesiveness of SBTI-treated cells to the internal abdominal surface suggested to them that proteases are involved in cell surface properties. Latner et al (41) and Latner and Turner (52) reported that aprotinin suppressed the growth of 2 different highly invasive carcinomas. The growth of Walker 256 carcinomas was inhibited in the rat peritoneum, but enhanced in rat lung by aprotinin (43). There was no effect on growth in rat muscle. The growth of a lymphosarcoma (44) was inhibited by aprotinin but not by EACA. Contradictory findings on the effects of aprotinin on Lewis lung tumors in mice were reported by Giraldi et al (44) and Turner and Weiss (46). The former found that aprotinin, but neither pepstatin nor leupeptin inhibited the growth and lung metastases of this tumor, while the latter enhanced metastases in the mouse lung. Giraldi et al (2) also observed that a natural leukocyte protease inhibitor caused significant reduction in growth and lung metastases in mice bearing Lewis lung carcinomas. Another natural protease inhibitor, isolated from calf cartilage, inhibited neovascularization of V2 carcinoma implanted in rabbit corneas (2). Aprotinin and FOY-305, a synthetic protease inhibitor, inhibited the growth of mouse skin squamous cell carcinoma induced with 3-MC (47,48). Greenbaum et al (49) observed that pepstatin inhibited the accumulation of ascites from 3 mammary tumors, 2 different leukemias and a sarcoma.

Table 3. Effects of protease inhibitors on growth of tumor cells in vivo.

Protease inhibitor	Effects/cell source/ host tissue[a]	References
SBTI	I/Ehrlich ascites/ mouse p.	(2),(40)
Aprotinin	I/transformed hamster kidney/hamster s.	(41)
Aprotinin	I/mouse mammary/mouse s.	(41),(42)
Aprotinin	NE/W/rat muscle	(43)
Aprotinin	E/W/rat lung	(43)
Aprotinin	E survival/W/rat p.	(43)
Aprotinin	I/mouse tumor/mouse s.	(44)
Pepstatin,leupeptin	NE/mouse tumor/mouse s.	(44)
Aprotinin	I/rodent lymphosarcoma/ rat s.	(45)
EACA	NE/rodent lymphosarcoma/ rat s.	(45)
Aprotinin	I lung metastasis/Lewis lung tumor/mouse s.	(46)
Aprotinin,FOY-305	I/3-MC-induced skin tumor/mouse skin	(47),(48)
Cartilage proteins	I of vascularization/V2 carcinoma/rabbit cornea	(2)
LNPI	I of growth + metastasis/ Lewis carcinoma/mouse s.	(2)
Pepstatin	I/various tumors/mouse p.	(49)

[a]W=Walker 256 carcinoma; s.=subcutis; p.=peritoneum. Other abbreviations as in Table 1.

4. Effects of protease inhibitors on properties of transformed and untransformed cells in vitro

Roblin (1) in an extensive review of the role of proteases on viral transformation summarized the effects of protease inhibitors on the properties of transformed and untransformed cells in vitro. A variety of protease inhibitors, with different modes of action and specificity produced inhibition of transformation-altered phenotypic properties. These included inhibition of cell division and proteolytic activity.

Table 4. Various effects of protease inhibitors on transformed and untransformed cells in vitro.

Protease inhibitor	Effects/cell system[a]	References
Leupeptin,antipain	I of SCE induced by PMA but not by X-rays/mouse	(53)
Antipain	I of MNNG-induced aberrations, but not SCE/V79 fibroblasts	(54)
Lima or kidney bean extract	E of RNA and DNA synthesis/lymphocytes	(51)
Bestatin	E of con A-stimulated DNA synthesis/lymphocytes	(52)
Leupeptin,pepstatin	I of differentiation/ embryonic rat pancreas	(55)
Leupeptin, ovomucoid	I of growth and DNA synthesis induced by DBN/frog bladder	(50)
Antipain, TLCK, leupeptin	I of Type C virus induction/Balb/c3T3	(56),(57)
Elastatinal	E of mutagenesis by MNNG/S.typhimurium	(2)
TLCK	I of growth and macromolecular synthesis/ mouse 3T3 and PY3T3	(2)
TPCK	I of peptide chain elongation/Hela,plasmocytoma, virus-transformed 3T3	(58)
Antipain,leupeptin	NE on DNA damage,repair or synthesis/human fibroblasts	(59)
Antipain	I of colony formation/ Xeroderma pigmentosum	(60)
Antipain	I of SOS functions/E. coli	(62)

[a]Abbreviations as in Table 1.

A summary of some effects published since this review was written (1975) are given in Table 4. Pietras (50) examined the alterations in growth and phenotypic properties produced in frog bladder cells by dibutylnitrosamine (DBN). This bladder carcinogen stimulated growth and DNA synthesis, release of cathepsin B1, and con A-mediated hemadsorption to the bladder cells. Leupeptin and ovomucoid, but not SBTI, inhibited the DBN-stimulated DNA synthesis and cell growth, and leupeptin reduced the DBN-stimulated increase in cathepsin B1 activity. Leupeptin, ovomucoid, TLCK, and iodoacetic acid, but not SBTI, inhibited the hemadsorption mediated by con A, while not affecting the binding of con A. Pietras remarks that only those agents which suppress cathepsin B1 activity altered this property.

The interpretation of the effects of chloromethylketone protease inhibitors such as TLCK and TPCK, which indicate a role for proteases in growth of tumor cells, should include their effects on SH-groups and metabolic processes as indicated by Roblin et al (1) and Rossman and Troll (2).

Contrary to most observations is the report of Harms-Ringdahl et al (51) that polypeptide trypsin inhibitors from lima or kidney beans stimulated RNA and DNA synthesis in lymphocytes, and Saito et al (52) found that bestatin, a small specific inhibitor of aminopeptidases, enhanced con A-stimulated DNA synthesis in these cells. These findings are not readily interpretable since proteolytic enzymes can also stimulate lymphocytes (52).

Contrary effects of leupeptin and antipain on sister chromatid exchange (SCE) were reported (53,54,59,60).

Some alterations in tissue development by protease inhibitors were observed. Marsh and Parsa (55) reported that leupeptin and antipain inhibited differentiation and morphogenesis in explanted embryonic rat pancreas tissue and Katz et al (61) found antipain and elastatinal to prevent normal mouse uterine maturation, which may be related to the inhibition of β-estradiol transformation (36).

A possible clue to the role of proteases in DNA functions

which lead to mutagenesis and carcinogenesis was provided by Meyn et al (62) who found that antipain inhibited the coordinate expression of diverse functions (SOS functions) that result from inhibition of DNA synthesis in E. coli by carcinogenic and mutagenic agents. These functions include prophage induction, filamentous growth, and an error-prone DNA repair activity that leads to mutagenesis. Antipain blocked UV mutagenesis in E. coli that possess rec $A^{\pm}$ and lex $A^{\pm}$ genes that are necessary for SOS functions. They stated that their results provided evidence for a key role of proteases in SOS functions. This was supported by Little et al (63) who showed that the rec $A^{\pm}$ protease specifically cleaves the lex $A^{\pm}$ protein. As Meyn et al stated "If an SOS-like repair function is responsible for carcinogenesis, it may require a proteolytic cleavage for induction or expression". Elastatinal was found (2) to suppress mutagenesis of S. typhimurium induced by MNNG and ENNG.

Viral induction in mammalian cells was also prevented by protease inhibitors. Long et al (56) showed that TLCK, antipain, and leupeptin inhibited Type C virus induction in Kirsten sarcoma virus-transformed Balb/c3T3 cells and Hellman et al (57) showed that antipain and leupeptin suppressed induction of Type C virus in another (A-12) virally-transformed Balb/c cell line. These studies indicate a role for proteases in mammalian cells that is similar to those found to operate in bacteria. It is of interest here that various tumor promoters in combined use with n-butyrate activated lymphoblastoid cells to express Epstein-Barr virus early antigen (58). The patterns of viral early polypeptides were similar from cells treated with n-butyrate alone or in combination with the promoters.

5. The role of proteases in the association of superoxide and peroxides with promotion

Recent developments have stimulated a growing interest in the possible role of activated oxygen compounds as ultimate mediators in tumor promotion (64) and carcinogenesis (22, 65-69). These developments include studies which implicate a significant role for proteases in the cellular production of activated oxygen species. These are summarized in Table 5.

Table 5. Association of active oxygen species with promotion and tumorigenesis: the role of proteases.

Agent	Effects/system[a]	References
PMA	SP/PMN,granulocyte	(22),(70-74)
Phorbol esters	SP by promoters only/PMN	(22)
Teleocidin,mezerein	SP/PMN	(22)
PMA	SP and HP, lipid peroxidation/macrophage, erythrocyte	(74)
PMA	DNA strand breaks associated with SP/leukocyte	(75)
PMA	Mitogenic and chromosomal damage suppressed by SOD/lymphocyte	(76)
PMA	SP and decreased SOD, no chromosomal damage/leukocyte	(77)
Various protease inhibitors	I of SP and HP by PMA/PMN	(22)
Various protease inhibitors	I of SP by con A and cytocholasin E/PMN,monocyte	(78)
TPCK	I of SP by FMLP/neutrophil	(79)
CuDIPS	I of SP by PMA/PMN	(80)
Protease inhibitors from legumes	I of SP by PMA/PMN	(81)
Organic peroxides,HP	Promoters/mouse skin	(82)
HP	Tumorigenic/mouse intestine	(83)
PMA, other promoters	Decreased SOD, catalase, lipid peroxidation, not affected by TPCK/mouse epidermis	(84),(85)
PMA	Decreased SOD, no effect on superoxide, no chromosomal damage/human fibroblasts	(77)
PMA	No effect on SOD, superoxide, chromosomes/V79	(77)
CuDIPS	I of PMA-induced promotion and ODC/mouse epidermis	(86)
CuDIPS	I of tumor growth/Ehrlich carcinomas in mouse p.	(87)

Table 5 (cont).

TLCK,TPCK	I of SP and oxygen consumption/PMA-stimulated rat macrophage	(88)

[a]HP=hydrogen peroxide; SP=superoxide production; FMLP= formyl methionyl leucyl phenylalanine; ODC = ornithine decarboxylase. Other abbreviations as in Table 1.

Biological oxidations, which often involve xanthine oxidase or NADPH-dependent oxidases, frequently cause the formation of superoxide (68,69) which is, for the most part, rapidly destroyed by 2 superoxide dismutases (SOD), a CuZnSOD present in the cytosol and mitochondria, and a MnSOD present in the mitochondria. The dismutase products include hydrogen peroxide and hydroxyl free radicals. These oxygen species are biologically destructive leading to DNA damage and lipid peroxidation.

PMA and other tumor promoters are potent stimulators of superoxide formation in white cells. Other cells have not yet been shown to release superoxide when treated with tumor promoters. The release of superoxide by white cells is believed to cause damage to neighboring cells.

Repine et al (70) were apparently the first to discover that small amounts (0.1-10 µg/ml) PMA stimulate polymorphonucleocytes (PMN) to undergo a phagocytic-type response, which includes an increase in oxygen uptake. This was confirmed by DeChatelet et al (71) who also observed that PMA stimulated the release of superoxide and hydrogen peroxide in PMN. The latter effect of PMA was repeatedly observed (2,68,69,73,74). Goldstein (2) suggested that the activation of PMN may be responsible for the inflammatory properties of PMA and other phorbol esters.

PMA, but not the non-promoter, 4-0-Methyl-PMA, stimulated the release of superoxide and hydrogen peroxide in macrophages where it also induced lipid peroxidation (74). Goldstein et al (2) found that the rate of superoxide production by PMN correlated with the tumor-promoting activity of various phorbol esters. Teleocidin B, a potent non-phorbol ester promoter was also effective. Mezerein, a weak promoter was more active than PMA. Human and rodent fibroblasts, however, were not stimulated to produce superoxide by PMA (89).

Direct evidence for the tumor-promoting and carcinogenic abilities of active oxygen compounds was recently obtained (82,83). Slaga et al (82) found that benzoyl peroxide is an effective skin promoter of papillomas and squamous cell carcinomas. Benzoyl peroxide was neither a tumor-initiator nor a complete carcinogen. Other properties of PMA were also exhibited by this agent (82), which is a free radical-generating compound. Other agents, which generate free radicals, such as chloroperbenzoic acid, lauroyl peroxide, and 2,2'-azobis-2-methylpropionitrile were also observed by this group to be effective skin tumor promoters. They showed that lauroyl peroxide was as effective as benzoyl peroxide in tumor promotion, while it also failed to act as an initiator or complete carcinogen. Hydrogen peroxide was found to be a very weak promoter but ITO et al (83) observed that it produced a small number of duodenal tumors in mice after oral administration.

A significant role for superoxide in tumor promotion and carcinogenesis is indicated by the studies of Leuthauser et al (87) and Kensler et al (86). They used the copper chelate CuDIPS, which is a low molecular weight ether soluble compound that exhibits SOD-like activity (80). This compound was injected im into mice carrying Ehrlich carcinoma cells (87). The tumor growth was decreased, the survival was increased, and lung metastases was delayed. The application of CuDIPS produced a marked inhibition of PMA promotion and PMA-stimulated ODC activity in mice (86). Kensler and Trush also observed (80) that CuDIPS as well as SOD suppressed the PMA-stimulated oxygen radical metabolism.

The metabolism of superoxide in mouse epidermis should be affected if it played an important role in tumor promotion. Solanki et al (84) found that mouse epidermal (SOD) and catalase were decreased by PMA and that SOD was decreased in papillomas. The promoters, anthralin, and phorbol dibutyrate, and the second stage promoter, mezerein, also lowered the activities of these enzymes, whereas the non-promoter, ethylphenyl propiolate, did not. Logani et al (85), however, observed that lipid peroxidation, an expected consequence of released active oxygen species, was decreased by PMA treatment. TPCK had no effect on this response.

We have also observed that PMA produces no effect on mouse epidermal lipid peroxidation after a single treatment (Table 6) but markedly suppresses lipid peroxidation after 6 treatments.

Table 6. Effects of single and multiple treatments of 10 μg PMA on malonaldehyde levels in mouse epidermis[a].

Hours after treatment	Malonaldehyde nmol/ μg DNA ($\pm$ S.E.)	
	Control	PMA
Single		
0	25.2 (3.5)	
10		15.2 (0.9)
18	15.4 (1.2)	12.2 (1.7)
24	23.0 (2.7)	12.2 (0.9)
27	15.0 (1.2)	11.0 (0.8)
30		15.6 (2.2)
33	27.6 (3.4)	21.0 (1.9)
40		16.6 (2.4)
Multiple[b]		
24	15.0 (1.7)	5.6 (0.6)[c]

[a] Each group consisted of 5 mice treated with 25 μg DMBA. PMA was applied 1 week after DMBA.

[b] These mice received PMA 3x/week for 2 weeks for a total of 6 treatments. Assays were done 24 hrs after last treatment.

[c] Significant at $p < 0.001$.

Stimulation of superoxide release by PMA has only been observed in white cells, as previously mentioned. This stimulation, however, may have promoting consequences by diffusion of superoxide to other target cells, where it may react with chromosomes. Birnboim (75) has shown that PMA produces significant DNA damage (single strand breaks) in leukocytes which was associated with superoxide production. This was inhibited by SOD and catalase. He has also found that benzoyl peroxide can directly produce DNA damage. Emerit and Cerutti (76) demonstrated that PMA induces chromosomal damage and SCE in PHA-stimulated lymphocytes. Both reactions were inhibited by SOD. The non-promoting phorbol and 4-0-Methyl-PMA were inactive. Kinsella et al (89), however, found that PMA did not induce superoxide production in human fibroblasts FF1 or Chinese hamster fibroblasts V79, although it did produce this response in human leukocytes. They also could not detect chromosomal aberrations in the lymphocytes. Decrease of SOD was obtained in lymphocytes and FF1 cells but not in V79. Repeated exposure to PMA over several passages did, however, result in elevated levels of chromosomal damage. No obvious reason can be given for the discrepancies between the observations by Emerit and Cerutti and those of Kinsella et al, although the presence of induced virus was a suggested (76, 78) possibility.

These important findings should be carefully repeated to determine whether PMA can indeed produce chromosomal damage via superoxide production. It would be interesting to also determine whether superoxide released by PMA-stimulation of white cells can damage fibroblasts or other target cells.

6. Role of protease

The role of proteases in superoxide production may be inferred from experiments that showed effects with proteases and protease inhibitors. Johnston et al (68) observed that trypsin, chymotrypsin, pronase, or papain caused macrophages to produce 2-6 times more superoxide in response to PMA. They do not report the effects of the proteases alone. The enzyme priming was blocked by protease inhibitors. Kitagawa et al (78) found various protease inhibitors to inhibit superoxide production in con A- or cyt E-stimulated white cells. The relative potencies for PMN were TPCK > TLCK > aprotinin > SBTI > phenylmethyl-sulfonylfluoride (PMSF); and TPCK > aprotinin > TLCK = SBTI > PMSF for monocytes.

Hoffman and Autor (88) observed that TLCK and TPCK suppressed oxygen consumption and superoxide production in PMA-stimulated rat pulmonary macrophages.

A positive feedback for protease activation may result from the ability of PMA-induced superoxide or hydrogen peroxide to destroy α1-protease or other protease inhibitors (77, 90).

The PMA-stimulated response of PMN, which include oxygen consumption, superoxide and hydrogen peroxide production was inhibited by various protease inhibitors (22). These inhibitors were SBTI, LBTI (lima), phosphoramidon, benzamidine, antipain, elastatinal, and chymotrypsin. The inactive protease inhibitors were pepstatin, EACA, leupeptin, TAME, trasylol, and a peptide chloromethyl ketone. These authors also reported (22) that SBTI suppressed superoxide production by a variety of stimulating agents. Superoxide production and other effects induced in neutrophils by PMA were inhibited by TPCK (79). Yavelow et al (81) showed that extracts from chick peas, kidney beans, and soy beans inhibited the PMA-stimulated production of superoxide in PMN. Purified protease inhibitors of trypsin and chymotrypsin from these legumes were also effective.

The review by Oberley et al (67) discusses the role of SOD in cancer. They report that there seems to be more evidence for a decrease in total tumor SOD as compared to normal tissue than there is for no change or increased activity. Since there are many cases of the latter, the lowered SOD in tumor cells is not a tumor characteristic, as some authors state. The review also mentions that lowered CuZnSOD is a common but not universal feature of tumors. The most significant characteristic appears to be a large decrease or absence of MnSOD. Their most meaningful conclusion was that the increase in net levels of superoxide is important and not the loss of MnSOD.

The release of superoxide may have other consequences for promotion and carcinogenesis in addition to DNA and other chromosomal damage. There is considerable evidence that it is a potent mediator of the inflammatory response which is associated with promoters and may be necessary for their action.

7. Role of Proteases in Promotion and Carcinogenesis

Although most recent work provides evidence (Table 7) that promoters and carcinogens stimulate, and tumor cells generate or maintain, increased amounts of proteases, the specific role of proteases in promotion and carcinogenesis has yet to be determined. Very little is known about the effect of these proteases on the initiation and maintenance of the tumorigenic state. Accumulating indications are that these enzymes are involved in mitogenesis and metastases.

A variety of proteases are present in homogenates of skin from humans and rodents (90). These include many exopeptidases, and proteases which resemble chymotrypsin, trypsin, and cathepsins with respect to substrates and inhibitors. Collagenases, elastases, and fibrinolysins are also present. Sulfhydryl-protease inhibitors which inhibited papain, ficin, and bromelain, but neither trypsin nor chymotrypsin were found in the cytosol or plasma membrane (92).

The function of proteases in mouse skin tumor promotion awaits the characterization of mouse epidermal proteases in normal epidermis and in that undergoing tumorigenesis. Shamberger and Rudolph (93) reported earlier that cathepsin was present in mouse skin cancers induced with 3-MC while no activity was observed in DMBA-initiated (but not promoted) skin.

Regulation of cell growth by proteases and their involvement in tumorigenesis was observed by Sefton and Rubin (1), Schnebli (1), and Schnebli and Burger (1). Sefton and Rubin reported that trypsin, pronase, and ficin stimulated DNA synthesis in confluent chick embryo cells and that trypsin stimulated mitosis and cell growth. SBTI inhibited the effects of trypsin but not of the other enzymes. Schnebli observed that transformed cells (PY3T3) contained several fold more protease activity than the normal parental cells (3T3). The protease was membrane-bound. The growth of PY3T3 was inhibited by TAME,

TLCK, and ovomucoid. Burger (1) also showed that proteases released normal fibroblasts from contact inhibition and induced the appearance of surface properties characteristic of transformed cells. The subsequent study by Schnebli and Burger (1) confirmed and extended these observations.

Table 7. Protease and promoter effects in normal and tumor cells.

Observations	Comments	References
Increased protease content or secretion in tumor cells	Cathepsin-like SH-proteases, collagenase, neutral protease, correlation with metastasis	(123-126), (133-137)
Increased secretion of fibrinolysins in tumor cells	Mostly plasminogen activators, correlation with metastatic potential	(107),(138), (139)
Increased secretion of anti-proteases	Anti-plasmin	(140)
Cell surface and secreted proteases increased in growing cells	Increased activity in some tumor cells	(107),(108), (140)
Mitogenic effect of proteases	Thrombin, fibroblasts	(142)
Induction of proteases by PMA and other promoters	Mostly plasminogen activator, elastase, collagenase, cathepsin, lysosomal enzymes	(110),(112), (116),(118), (121),(122), (127-130),(139), (142-144)
Protease inhibitors inhibit PMA-induced protease	Plasminogen activator in transformed fibroblasts; chymostatin most active	(39)
Protease inhibitors inhibit PMA and RA-induced transglutaminase	Mouse epidermal cells in vitro; chymostatin most active	(146)

A direct role for protease-stimulation of DNA synthesis was indicated by Brown et al (94) who found that trypsin activated G1, but not S phase, nuclei from Chinese hamster fibroblasts, to synthesize DNA. Trypsin caused an increase in DNA polymerase activity which provided a specific mechanism for protease-initiation of DNA synthesis.

Proteases appear to be mediators in three processes that seem to be important in promotion and tumorigenesis. These are: a) inflammation, b) cell growth, and c) metastases. The evidence for promoter-stimulation of these processes, with regard to protease mediation, was previously reviewed (2). More recent and additional evidence will be considered.

It is of some interest that natural protease inhibitors are elevated in the urine and blood of cancer patients. These are α1-antitrypsin (95,96) and α1-antichymotrypsin (97,98). An α1-antichymotrypsin-like protein which was immunologically identical to the human protein, but devoid of anti-chymotryptic activity was found in nude mice carrying a human malignant melanoma (99).

The inflammatory response associated with tumor promoters involves white cells (100) and possibly platelets (100) whose stimulation causes the release of biochemical mediators that include histamine (100), active oxygen species, proteases such as collagenase, elastase, and cathepsins, and the products of the arachidonic acid cascade (100). The role of proteases in promoter-stimulated release of oxygen species was discussed above.

Arachidonic acid metabolism produces two series of metabolites generated by cyclooxygenase and lipoxygenase. The former causes the release of various prostaglandins and thromboxanes, whereas the latter is responsible for the formation of various leukotrienes as well as hydroxy and hydroperoxy derivatives of eicosatetraenoic acid. These metabolites appear to be involved in mechanisms of inflammation, cell growth and differentiation, and growth and metastases of tumors (101,102).

The effects of tumor promoters in cellular proliferation, protease induction, and promotion, seem to involve the metabolites of arachidonic acid (101-105). Some of these have also been shown to be mediated by proteases. Levine (104) observed that leupeptin inhibited the PMA-stimulated arachidonic acid metabolism of smooth muscle cells. Crutchley and Maynard (145) found that lipoxygenase, but not cyclooxygenase inhibitors inhibited the induction of plasminogen activator by PMA. Chang et al (106) reported that PMA and various proteolytic enzymes induced the synthesis of PGE_2 in mouse peritoneal macrophages. A variety of protease inhibitors inhibited both protease- and PMA-induced PGE_2 synthesis.

PMA and other tumor promoters induce protease release from many cells (Table 7). Plasminogen activator has been more extensively studied than the others. Cell surface protease activity, which is not plasminogen activator, has been observed by Hatcher (107,108) to be correlated with doubling time of cells.

They also found that transformed cells have more proteolytic activity than normal ones. Plasminogen activator and its product, plasmin, (109-112), along with collagenase (110), cathepsin B (113,114), and elastase (115) are considered to be involved in neoplasia and metastases. Reich (111) makes the strongest claim for the association between neoplasia and plasminogen activator. His emphasis is on data obtained from primary malignancies in man and animals and primary cultures of cells. He presents evidence for an elevated activator production as a malignant phenotype. Reich also provides evidence for a strong association between tumor promoters and their ability to induce high levels of plasminogen activator.

Mott et al (116) reported a lack of correlation between fibrinolysis and transformation of ten extensively cultured mammalian cells, which are not considered by Reich to be adequate models for cancer studies. Plagemann and Estensen (117) have also observed a lack of correlation between various normal and transformed cells in culture. In addition, only one of their cells was stimulated to produce plasminogen activator by PMA. They also point out that their cells may already be maximally stimulated.

Long et al (112) demonstrated that antipain, but neither leupeptin nor SBTI suppressed the PMA-stimulated induction of PA activity in normal 10T½ cells. They also observed that only leupeptin inhibited the PA production in UV or X-ray transformed cells. The various effects of protease inhibitors of transformation of 10T½ cells (36) indicated that different types of proteases exert specific actions which are affected by the various protease inhibitors.

Brynes et al (118) compared a series of phorbol derivatives for their ability to induce PA in chick embryo fibroblasts, aggregate platelets, and act as irritants and promoters. They obtained very good correlation between PA, skin irritancy, and promotion and a poorer correlation with platelet aggregation and promotion.

A converse effect on PA induction is its inhibition by PMA in myeloid leukemia cells (119). These cells are induced to differentiate by PMA which inhibits differentiation in Friend erythroleukemia cells (120). No measurements of PMA-inducibility of PA in the Friend cells were made. The association between the ability of PMA to induce PA and its ability to induce or inhibit differentiation would be of great interest. There are eleven cell systems whose differentiation is inhibited and four cell systems whose differentiation is induced by PMA. Thus far, PMA has been shown to induce PA in 3T3 (121) and chick embryo fibroblasts (118, 122) whose differentiation is inhibited by PMA.

Chymostatin was the most effective protease inhibitor to inhibit the PMA-stimulated release of PA from transformed fibroblasts (39) as well as the PMA and retinoic (RA)-induction of transglutaminase in epidermal cells (146). The induction of epidermal differentiation which involves transglutaminase activity may be an important component of promotion. It was noted (146) that the possible role of proteases in differentiation could account for the antipromoting action of protease inhibitors.

The induction of PA may be specific for phorbol ester-type promoters. Wigler et al (122) found that several other type promoters, such as anthralin, Tween 60, and asbestos, did not induce PA in Hela or chick embryo fibroblast cells.

The general role of PA and other proteases should be assessed by determination of the ability of these other promoters to induce proteases in target cells and by determination of the ability of protease inhibitors to inhibit promotion by these agents. Similar studies in other tissues promoted by PMA (120) would also contribute to an appreciation of the role of proteases in promotion.

Other proteases stimulated by PMA have been associated with neoplasia or metastases. Cathepsin B, found by Dolbeare (123) to be induced in 10T½ cells by promoters, is secreted by mammary carcinomas (114,124).

Ehrlich ascites cells (125) have a cell surface trypsin-like protease which activates a collagenase zymogen secreted by these cells. This protease is inhibited by TLCK but not by SBTI.

Collagenases and elastase are additional enzymes presumed to be involved in metastases. Tarin et al (126), for example, has obtained evidence that shows a correlation of collagenase secretion with metastatic potential in mouse mammary tumors.

PMA stimulated rabbit synovial fibroblasts to produce large amounts of collagenase (127,128). Brinckerhoff et al (129) found that PMA induces translatable collagenase mRNA in these cells (129). Moscatelli et al (110) found PMA to stimulate latent collagenase production in cultured human umbilical cord endothelial cells. They also found that the non-promoting esters, 4-0-Me-PMA and 4-α-PDD, were inactive. The collagenase activity was not inhibited by antipain, SBTI, pepstatin or other protease inhibitors.

Dahlgren et al (130) observed that PMA induced the secretion of elastase in mouse peritoneal macrophages. Many protease inhibitors were tested but no inhibition was obtained with the usual agents.

Platelet aggregation is another factor that appears to be associated with metastases (114). PMA was found to be the most potent aggregating agent (131). TLCK and TPCK inhibited both PMA and ADP-induced aggregation. Thrombin is a well known stimulator of platelet aggregation, which was inhibited by DAPA, a synthetic specific thrombin inhibitor (132).

Cathepsin B activity correlates with metastatic potential and platelet aggregation (113,114). Honn et al (114) found that papain, which resembles cathepsin B, is a potent platelet aggregating agent. Leupeptin, antipain, and TLCK, inhibited platelet aggregation, but SBTI and aprotinin were weak or inactive inhibitors. They also observed that leupeptin, antipain, and chymostatin were potent inhibitors of cysteine proteinase activity of B16 melanoma cells.

C. SUMMARY

The piebald protean panoply of protease associated promoter-produced processes may reflect the processional program that prevails in tumorigenesis. Each of the sequential stages that lead to tumor formation sequesters several scheduled switches that successively set the progressional events that culminate in the conversion of normal to tumor cells. It is plausible that one or more specific proteases operate at several of these events.

The determination of where and how specific proteases act might be achieved by focusing on likely areas amenable for exploration. One possible site is the plasma membrane, which contains proteases, and which is believed to be the specific and major target of PMA (120) and possibly of proteases as well. Proteases may be tested as cocarcinogens, initiators and promoters in cell transformation assays.

Recent work of Slaga et al (13) shows that promotion is composed of at least two stages and that TPCK inhibits Stage I and not Stage II. Although this implicates protease action in the first but not the latter stage, it is desirable that other types of protease inhibitors be tested, such as leupeptin and chymostatin. In general the use of more specific types of protease inhibitors is needed to determine the role of proteases in promotion.

II. CYCLIC NUCLEOTIDES

The role of cyclic nucleotides in the biochemical mechanisms involved with tumor promotion has been extensively examined by a number of laboratories during the past decade. Although many questions have not yet been satisfactorily answered and, in some cases, conflicting results have been published, several general conclusions may be reached from the body of accumulated data.

The literature on the cyclic nucleotides is vast, including a number of excellent reviews. Aspects of the field that are relevant to tumor promotion include the relationship of cyclic AMP and cyclic GMP to growth control (147-149), the role of cyclic AMP as a second messenger for membrane active agents (150,151), the relationship between prostaglandins and hormones to cAMP levels (151-153), and the study of cAMP dependent protein kinases (154-156). There is also considerable literature of cyclic nucleotides and cancer (157).

The identification of phorbol myristate acetate (PMA) as the active tumor promoting agent in croton oil (158,159) and the elucidation of the lipophilic chemical structure of the promoter, led to the idea that the plasma membrane might serve as the initial or primary target site of action (160,161). This would imply some sort of transduction signal from the plasma membrane to account for the pleiotypic cellular biochemical response to PMA. For example, the induction of ornithine decarboxylase by PMA (162,163) requires gene activation and protein synthesis. Cyclic AMP has been implicated as a second messenger in the induction of ornithine decarboxylase in other systems (164), and the possibility that PMA acts via cAMP to induce ornithine decarboxylase in mouse epidermis was investigated (165). The results demonstrated that neither cAMP nor cGMP were involved in this PMA effect. A similar conclusion was reached in studies in Chinese hamster ovary cells and mutants lacking cAMP dependent protein kinase (166,167).

Although early work indicated that basal cAMP levels were affected by PMA treatment in mouse epidermis (168), this was later shown to be an artifact of epidermal preparations (169). When epidermis was removed from whole skin in a way that avoids ischaemia (170,171) no effect on basal cAMP was seen between 2 and 72 hours (172). However, a large (3-7 fold) increase in cAMP was seen between 10 minutes and 2 hours after PMA treatment (173). Increases in cGMP levels have been found more consistently at several time points after a single application of PMA to mouse skin (172,173).

The ratio of cGMP to cAMP in individual samples was increased from 36 to 72 hours after a single topical PMA treatment (172). Multiple treatments of PMA results in a large (5-10 fold) persistent elevation of cGMP (174). This phenomenon shows a dose response relationship with no effect seen at doses of less than 1 μg PMA (175). In a study comparing several mouse strains (shown in Table 1), no correlation was found between the elevation of cGMP levels in response to multiple PMA treatments and sensitivity to tumor promotion (175). These results cast considerable doubt on the primacy of the cGMP elevation response as a critical step in the biochemical mechanism of tumor promotion.

Table 8. PMA Effects on c-GMP Levels in Various Mouse Strains.

Mouse Strain	Relative Sensitivity	cGMP[a]		
		Control[c]	PMA[d]	PMA/cont.
SENCAR	++++	6.66 ± 0.62	8.70 ± 2.72	1.31
Ha/ICR	+++	2.57 ± 0.19	1.88 ± 0.25	0.73
CD1	+++	6.70 ± 1.28	16.2 ± 1.69	2.42
Balb	++	6.43 ± 0.33	5.68 ± 0.98	0.88
C57B1	+	6.47 ± 1.07	16.1 ± 3.18	2.49

[a] fmole/μg DNA ± SEM - assayed by RIA.
[b] Relative sensitivity to skin carcinogenesis by initiation-promotion with PMA based on literature.
[c] Treated 2X with 0.2 ml acetone.
[d] Treated 2X with 0.2 ml acetone containing 10 μg PMA.

Other cell systems have been used to study the effects of PMA on cyclic nucleotide levels. Large, rapid and transient alterations in cGMP and/or cAMP levels were observed in chick myoblasts (176), human lymphocytes (177), rat embryo fibroblasts (178), mouse fibroblasts (179), and human platelets (179). These observations are potentially interesting with respect to the mitogenic action of phorbol ester, but do not significantly address the issue of the role such changes might play in mouse skin two stage carcinogenesis.

Grimm and Marks (180) first reported the inhibition of isoproterenol induced accumulation of cAMP in mouse epidermis. This effect has been consistently observed in several laboratories (169,181,182). A similar decreased β-adrenergic response after PMA treatment was seen in glial tumor cells (183) and rat fibroblasts (184). The loss of β-adrenergic response following treatment of mouse epidermis in vivo to PMA is not due to loss of β-receptor number or function (182) but to a decrease in coupling efficiency of the receptor to adenyl cyclase (185). The effect becomes an irreversible phenotypic change in papillomas arising during promotion with PMA (186). Although a recent publication (187) suggested that superoxide production by PMA stimulated macrophages may be responsible for the loss of β-adrenergic response in promoter treated mouse skin, we have recently demonstrated the same inhibition by PMA in cultured epidermal cells in vitro (188), a finding which supports some direct action of the promoter on the target cell membrane. In the original paper of Grimm and Marks (130) a correlation was seen between the promoting activity of a number of compounds and the inhibition of β-adrenergic response. However, these authors have more recently shown that other non-promoting phorbol analogs such as 12-retinoyl phorbol acetate (RPA) also produce the same effect (189). Mufson et al. (190) had previously shown loss of β-adrenergic response by mezerein, another non-promoting agent with many biochemical similarities to phorbol ester. These data tend to put the uncoupling of β-adrenergic receptors by PMA into the large category of biochemical effects of PMA which may be necessary but not sufficient for expression of tumor promoting activity. An alternate possibility is that uncoupling of β-adrenergic receptor is a biochemical parameter of the second stage of tumor promotion (191), since both RPA and mezerein are second stage promoters (191,192).

Rochette-Egly and Castagna (193) found that prostaglandin (PG) El stimulated cAMP accumulation in rat embryo fibroblasts was inhibited by PMA, as well as was catecholamine stimulation. This observation is of considerable potential significance because of subsequent data from a number of laboratories that PMA stimulates prostaglandin production in mouse epidermis (194-196). An uncoupling action operating on the prostaglandin receptor-adenyl cyclase system would certainly have some bearing on the relationship of PMA stimulated PG to subsequent events in the mechanism of tumor promotion (197). We therefore decided to determine the effect of PMA on PG stimulated cAMP accumulation in mouse epidermis. We found, however, that mouse epidermis *in vivo*, epidermal explants *in vitro*, and a mouse epidermal cell line were refractory to cAMP stimulation by a 100-fold concentration range of either PGE1 or PGE2 (198). A similar result was found *in vivo* by Marks *et al*. (199). Mouse epidermis is therefore one of the few tissues which do not respond to PGE with increased cAMP levels (200-202).

An important mechanism by which cAMP functions as a second messenger is the activation of cAMP-dependent protein kinases. The possibility that certain kinases may be involved in the action of viral and cellular oncogenes (203), and the recent finding that the PMA receptor is a protein kinase (204,205) lend importance to studies on the effects of PMA on the cAMP dependent kinases. In B16 mouse melanoma cells, PMA suppressed cAMP-dependent kinase activity while not directly affecting basal or hormone stimulated levels of cAMP (206). In mouse epidermis, however, no change was seen in either cAMP dependent or independent kinase activities after PMA treatment *in vivo* (207).

The studies discussed so far relate to the effects of tumor promoting agents on endogenous cyclic nucleotide metabolism. An alternative approach to the issue of the role of cyclic nucleotides in tumor promotion is to examine the effects of these molecules on the carcinogenic process. Curtis et al. (208) found inhibition of two stage carcinogenesis by injection of cAMP simultaneously with promoter treatment. These results were confirmed and extended by

Perchollet and Boutwell (209) who found inhibition of tumor promotion by topical application of cAMP, or by an inhibitor of cAMP phosphodiesterase, 3-isobutyl-1-methylxanthine. Theophyline, another phosphodiesterase inhibitor, was earlier shown to inhibit tumor promotion (168).

It is clear by now that the fundamental biochemical mechanisms associated with tumor promotion are complex. A major challenge has been to separate critical causative phenomena, from secondary and resultant effects. The demonstration by Slaga et al. (191) of multiple stages of promotion, each presumably involving different biochemical mechanisms, increases the complexity of interpretation. Although no clear evidence has been found for a direct and causative role of either cyclic nucleotide in the complete mechanism of tumor promotion, certain conclusions may be safely drawn from the data generated to date. The cyclic nucleotides do not appear to play a second messenger function in triggering specific activities stimulated by PMA. It is unlikely that cyclic nucleotides have any role in the first stage of promotion. It seems highly probable that the maintenance of a low level of cellular cAMP may be necessary for the second stage of tumor promotion. This is consistent with the inhibitory effects of cAMP on tumor promotion and the prevention by second stage promoters including PMA of hormonal stimulation of adenyl cyclase. It is important to note that the tumor promoter PMA may exert disparate effects on different cell types. For example, in mouse myeloid leukemia cells, PMA treatment results in enhanced isoproterenol response (210), while in turkey erythrocytes no effect is seen (187). Effects of PMA on non-epidermal cells may be valuable as probes of various cellular processes, including mitogenesis and regulatory control, but considerable caution should be used in the extrapolation of these effects to mechanisms of tumor promotion in mouse skin.

ACKNOWLEDGEMENTS

We are grateful to Dr. W. Troll for a critical review of the protease section.

References

1. Roblin R, Chow L, Black PH: Proteolytic enzymes, cell surface changes, and viral transformation. Adv Cancer Res (22): 203-260, 1975.
2. Rossman TG, Troll W: Protease inhibitors in carcinogenesis: possible sites of action. In: Slaga TJ (ed) Carcinogenesis Vol. 5, Modifiers of chemical carcinogenesis. Raven Press, New York, 1980, pp 127-143.
3. Ribbons DW, Brew K (eds) Proteolysis and Physiological regulation. Academic Press, New York, 1976.
4. Clarkson B, Baserga R (eds) Control of proliferation in animal cells. Cold Spring Harbor Laboratory, 1974.
5. Reich E. Rifkin, DB, Shaw E (eds) Proteases and biological control. Cold Spring Laboratory, 1975.
6. Holzer H, Tschesche H (eds) Biological functions of proteinases. Springer-Verlag, New York, 1979.
7. Barrett AJ (ed) Proteinases in mammalian cells and tissues. North-Holland, New York, 1979.
8. Fritz H, Tschesche H, Greene LJ, Truscheit E (eds) Proteinase inhibitors. Springer-Verlag, New York 1974.
9. Troll W, Klassen A, Janoff A: Tumorigenesis in mouse skin: Inhibition by synthetic inhibitors of proteases. Science (169): 1211-1213, 1970.
10. Matsushima T, Kakizoe T, Kawachi T, Hara K, Sugimura T, Takeuchi T, Umezawa H: Effects of protease inhibitors of microbial origin on experimental carcinogenesis. In: Magee PN, Takayama S, Sugimura T, Matsushima T (eds) Fundamentals in cancer prevention. University Park Press, Baltimore, 1976, pp 57-69.
11. Janoff A, Klassen A, Troll W: Local vascular changes induced by the cocarcinogen, phorbol myristate acetate. Cancer Res (30): 2568-2671, 1970.
12. Belman S: Unpublished observation.
13. Slaga TJ, Klein-Szanto AJP, Fischer SM, Weeks CE, Nelson K, Major S: Studies on mechanism of action of anti-tumor-promoting agents: Their specificity in two-stage promotion. Proc Natl Acad Sci USA (77): 2251-2254, 1980.

14. Goto M, Iguchi Y, Ozawa H, Sato H: Change of polyamine content in mouse skin by leupeptin, a protease inhibitor, during early stage of tumorigenesis. Gann (71): 18-23, 1980.
15. Kakizoe T, Esumi H, Kawachi T, Sugimura T, Takeuchi T, Umezawa H: Further studies on the effect of leupeptin, a protease inhibitor, on induction of bladder tumors in rats by N-butyl-N-(4-hydroxybutyl)nitrosamine. J Natl Cancer Inst (59): 1503-1508, 1977.
16. Kakizoe T, Matsukura N, Aoyagi M, Kawachi T, Sugimura T: Effect of simultaneous administration of leupeptin on induction of bladder tumors in rats by (N-butyl-N-4-hydroxybutyl)nitrosamine. Gann (71): 138-140, 1980.
17. Kakizoe T, Sano T, Kawachi T, Sugimura T, Takeuchi T, Umezawa H: Effect of leupeptin on induction of lymphoblastic leukemia in mice by N-nitrosobutylurea. Gann (68): 282-285, 1977.
18. Yamamura M, Nakamura M, Fukui Y, Takamura C, Yamamoto M, Minato Y, Tamura Y, Fuji S: Inhibition of 7,12-dimethylbenz a anthracene-induced mammary tumorigenesis in rats by a synthetic protease inhibitor, [N,N-dimethylamino-p-(p'-guanidobenzoyloxyl) benzilcarbonyloxy] glycolate. Gann (69): 749-752, 1978.
19. Corasanti JG, Hobika GH, Markus G: Interference with dimethylhydrazine induction of colon tumors in mice by ε-aminocaproic acid. Science (216): 1020-1021, 1982.
20. Hosaka S, Hirono I: Effect of leupeptin, a protease inhibitor, on the development of spontaneous tumors in strain A mice. Gann (71): 913-917, 1980.
21. Becker FF: Inhibition of spontaneous hepatocarcinogenesis in C3H/HeN mice by Edi Pro A, an isolated soy protein. Carcinogenesis (2): 1213-1214, 1981.
22. Troll W, Witz G, Goldstein B, Stone D, Sugimura T: The role of free oxygen radicals in tumor promotion and carcinogenesis. In: Hecker E, Kunz W, Fusenig NE, Marks F, Thielmann Hw (eds) Cocarcinogenesis and biological effects of tumor promoters. Carcinogenesis Vol. 7, Raven

Press, New York, 1982, pp 593-597.
23. Barrett AJ, Kembhavi AA, Brown MA, Kirschke H, Knight CG, Tamai M, Hanada K: L-<u>trans</u>-Epoxysuccinyl-leucylamido (4-guanidino)butane (E-64) and its analogues as inhibitors of cysteine proteinases including cathepsins B, H, and L. Biochem J (201): 189-198, 1982.
24. Brown WE, Wold F: Alkyl isocyanates as active-site-specific reagents for serine proteases. Reactive properties. Biochem (121): 828-840, 1973.
25. Beeley JG, Neurath H: The reaction of trypsin with bromoacetone. Biochem (71): 1239-1251, 1968.
26. Husain SS, Lowe G: The location of the active-site histidine residue in the primary sequence of papain. Biochem J (108): 861-866, 1968.
27. Husain SS, Lowe G: Evidence for histidine in the active site of papain. Biochem J (108): 855-859, 1968.
28. Husain, SS, Lowe G: Evidence for histidine in the active sites of ficin and stem-bromelain. Biochem J (110): 53-57, 1968.
29. Yavelow J, Finley TH, Kennedy AR, Troll W: Bowman-Birk soybean protease inhibitor as an anticarcinogen. Cancer Res (43): 2454s-2459s, 1983.
30. Dermer GB: Human cancer research. Science (221): 318, 1983.
31. Borek C, Miller R, Pain C, Troll W: Conditions for inhibiting and enhancing effects of the protease inhibitor antipain on X-ray-induced neoplastic transformation in hamster and mouse cells. Proc Natl Acad Sci USA (76): 1800-1803, 1979.
32. Kennedy AR, Little JB: Effects of protease inhibitors on radiation transformation <u>in</u> <u>vitro</u>. Cancer Res (41): 2103-2108, 1981.
33. Beard CR, Rutledge-Freeman N, Miller RC, Borek C: Antipain and radiation effects on oncogenic transformation and sister chromatid exchange in Syrian hamster embryo and mouse C3H/10T1/2 cells. Carcinogenesis (2): 1229-1233, 1981.

34. Di Paolo JA, Amsbaugh SC, Popescu NC: Antipain inhibits N-methyl-N'-nitro-N-nitrosoguanidine-induced transformation and increases chromosomal aberrations. Proc Natl Acad Sci USA (77): 6649-6653, 1980.
35. Kuroki T, Devron C: Inhibition of chemical transformation in C3H/10T1/2 cells by protease inhibitors. Cancer Res (39): 2755-2761, 1979.
36. Kennedy AR, Weichselbaum RR: Effects of 17β-estradiol on radiation transformation in vitro: inhibition of effects of protease inhibitors. Carcinogenesis (2): 67-69, 1981.
37. Kennedy A: Personal communication.
38. Kennedy AR, Little JB: Protease inhibitors suppress radiation-induced malignant transformation in vivo. Nature (276): 825-826, 1978.
39. O'Donnell-Tormey J, Quigley JP: Inhibition of plasminogen activator release from transformed chicken fibroblasts by a protease inhibitor. Cell (27): 85-95, 1981.
40. Whur P, Robson RT, Payne NE: Effect of a protease inhibitor on the adhesion of Ehrlich ascites cells to host cells in vivo. Br J Cancer (28): 417-428, 1973.
41. Latner AL, Longstaff E, Turner GA: Anti-tumor activity of aprotinin. Br J Cancer (30): 60-67, 1974.
42. Latner AL, Turner GA: Effect of aprotinin on immunological resistance in tumour-bearing animals. Br J Cancer (33): 535-538, 1976.
43. Thompson AW, Pugh-Humphries RGP, Horne CHW, Tweedie DJ: Aprotinin and growth of Walker 256 carcinosarcoma in the rat. Br J Cancer (35): 454-460, 1977.
44. Giraldi T, Nisi C, Sava G: Lysosomal enzyme inhibitors and antimetastatic activity in the mouse. Eur J Cancer (13): 1321-1323, 1977.
45. Back N, Steger R: Effect of aprotinin, EACA, and heparin on growth and vasopeptide system of Murphy-Sturm lymphosarcoma. Eur J Pharmacol (38): 313-319, 1976.
46. Turner GA, Weiss L: Analysis of aprotinin-induced enhancement of Lewis lung tumors in mice. Cancer Res (41): 2576-2580, 1981

47. Ohkoshi M, Fuji S: Effect of oral administration of protease inhibitor [N,N-dimethyl-carbamoylmethyl 4-(4-guanidinobenzoyloxyl)-phenylacetate] methanesulfonate on the growth of 3-methylcholanthrene-induced carcinomas in mice. Gann (73): 108-110, 1982.
48. Ohkoshi M: Effect of aprotinin on growth of 3-methylcholanthrene-induced squamous cell carcinoma in mice. Gann (71): 246-250, 1980.
49. Greenbaum LM, Esumi H, Sato S: Further studies of the effect of pepstatin on ascites accumulation in tumor-bearing mice. Cancer Lett (7): 91-96, 1979.
50. Pietras RJ: Heritable membrane alterations and growth associated with enhanced leupeptin-sensitive proteinase activity in epithelial cells exposed to dibutylnitrosamine in vitro. Cancer Res (38): 1019-1030, 1978.
51. Harms-Ringdahl M, Forsberg J, Fedorcsak I, Ehrenberg L: Trypsin inhibitory activity of a polypeptide isolated from red kidney beans, that also enhances lymphocyte stimulation. Biochem Biophys Res Commun (86): 492-499, 1979.
52. Saito M, Aoyagi T, Umezawa H, Nagai Y: Bestatin, a new specific inhibitor of aminopeptidases, enhances activation of small lymphocytes by concanavalin A. Biochem Biophys Res Commun (76): 526-533, 1977.
53. Little JB, Nagasawa H, Kennedy AR: DNA repair and malignant transformation: effect of X-irradiation, 12-0-tetradecanoyl-phorbol-13-acetate, and protease inhibitors on transformation and sister-chromatid exchange in mouse 10T1/2 cells. Radiat Res (79): 241-255, 1979.
54. Kinsella AR, Radman M: Inhibition of carcinogen-induced chromosomal aberrations by an anticarcinogenic protease inhibitor. Proc Natl Acad Sci USA (77): 3544-3547, 1980.
55. Marsh WH, Parsa I: Antiproteases and arrest of pancreatic differentiation. Fed Proc (37): 1699, 1978.
56. Long CW, Bruszewski JA, Christensen WL, Suk WA: Effects of protease inhibitors on chemical induction of Type C

virus. Cancer Res (39): 2995-2999, 1979.
57. Hellman KB, Brewer PP, Twardzik DR, Hellman A: Protease inhibitors modify induction of endogenous Type C oncornavirus (41019). Proc Soc Exp Biol Med (166): 28-34 1981.
58. Kawanishi M, Ito Y: Similarity of Epstein-Barr virus early polypeptides induced by various tumor promoters. Cancer Lett (16): 18-23, 1982.
59. Borek C, Cleaver JE: Protease inhibitors neither damage DNA nor interfere with DNA repair or replication in human cells. Mutat Res (82): 376-380, 1981.
60. Takeda H, Ishizaki K: High sensitivity of Xeroderma Pigmentosum cells to antipain, a protease inhibitor. Proc Japan Cancer Assoc (11): 13, 1978.
61. Katz J, Troll W, Genunchi A, Levitz M: Prevention of normal mouse uterine maturation by antipain and elastatinal: suppression of peroxidase activity. Endocrine Res (5): 325-335, 1978.
62. Meyn MS, Rossman T, Troll W: A protease inhibitor blocks SOS functions in Escherichia Coli: antipain prevents repressor inactivation, ultraviolet mutagenesis, and filamentous growth. Proc Natl Acad Sci USA (74): 1152-1156, 1977.
63. Little JW, Edmiston SH, Pacelli LZ, Mount DW: Cleavage of the Escherichia coli lexA protein by the recA protease. Proc Natl Acad Sci USA (77): 3225-3229, 1980.
64. Marx JL: Do tumor promoters affect DNA after all? Science (219): 158-159, 1983.
65. Oberley LW, Oberley TD, Buettner GR: Cell division in normal and transformed cells: the possible role of superoxide and hydrogen peroxide. Med Hypotheses (7): 21-42, 1981.
66. Totter JR: Spontaneous cancer and its possible relationship to oxygen metabolism. Proc Natl Acad Sci USA (77): 1763-1767, 1980.
67. Oberley LW, Buettner GR: Role of superoxide dismutase in cancer: a review. Cancer Res. (39): 1141-1149, 1979.

68. Bannister JV, Hill HAO (eds) Chemical and biochemical aspects of superoxide and superoxide dismutase. Developments in Biochemistry, Vol 11A, Elsevier/North Holland, New York, 1980.
69. Bannister WH, Bannister JV (eds) Biological and clinical aspects of superoxide and superoxide dismutase. Developments in Biochemistry, Vol. 11B, Elsevier/North Holland, New York, 1980.
70. Repine JE, White JG, Clawson CC, Holmes BM: Effects of phorbol myristate acetate on the metabolism and ultrastructure of neutrophils in chronic granulomatous disease. J Clin Invest (54): 83-90, 1974.
71. DeChatelet LR, Shirley PS, Johnston, Jr., RB: Effect of phorbol myristate on the oxidative metabolism of human polymorphonuclear leukocytes. Blood (47): 545-554, 1976.
72. Golstein IM: Effects of phorbol esters on polymorphonuclear leukocyte functions in vitro. In: Slaga TJ, Sivak A, Boutwell RK (eds) Carcinogenesis, Vol. 2. Mechanisms of tumor promotion and cocarcinogenesis. Raven Press, New York, 1978, pp 389-400.
73. Badway JA, Curnutte JT, Robinson JM, Lazduns JK, Briggs RT, Karnovsky MJ, Karnovsky ML: Comparative aspects of oxidative metabolism of neutrophils from human blood and guinea pig peritonea: magnitude of the respiratory burst, dependence upon stimulating agents, and localization of the oxidases. J Cell Physiol (105): 541-551, 1980.
74. Pick E, Keisari Y, Bromberg Y, Freund M, Yakerbowski A: Effect of tumor promoters in immunological systems-The macrophage as a target cell for the action of phorbol esters. In Hecker E, Kunz W, Fusenig NE, Marks F, Thielmann HW, (eds) Cocarcinogenesis and biological effects of tumor promoters, cocarcinogenesis, a comprehensive survey, Vol. 7, Raven Press, New York, 1982, pp 625-635.
75. Birnboim HC: DNA strand breakage in human leukocytes exposed to a tumor promoter, phorbol myristate acetate. Science (215): 1247-1249, 1982.

76. Emerit I, Cerutti PA: Tumour promoter phorbol-12-myristate-13-acetate induces chromosomal damage via indirect action. Nature (293): 144-146, 1981.
77. Clark RA, Stone PJ, Hag AE, Calore JD, Franzblow C: Myloperoxidase-catalyzed inactivation of al - protease inhibitor by human neutrophils. J Biol Chem (256): 3348-3353, 1981.
78. Kitagawa S, Takaku F, Sukamoto S: Evidence that proteases are involved in superoxide production by human polymorphonuclear leukocytes and monocytes. J Clin Invest (65): 74-81, 1980.
79. Duque RF, Phan SH, Sulavik MC, Ward PA: Effect of protease inhibitors on depolarization of the transmembrane potential in rat neutrophils. Correlation with superoxide generation and enzyme release. Fed Proc (42): 385, 1983.
80. Kensler TW, Trush MA: Inhibition of oxygen radical metabolism in phorbol ester-activated polymorphonuclear leukocytes by an antitumor promoting copper complex with superoxide dismutase-mimetic activity. Biochem Pharmacol in press.
81. Yavelow J, Gidlund M, Troll W: Protease inhibitors from processed legumes effectively inhibit superoxide generation in response to TPA. Carcinogenesis (3): 135-138, 1982.
82. Slaga TJ, Klein-Szanto AJP, Triplett LL, Totti LP, Trosko JE: Skin tumor-promoting activity of benzoyl peroxide, a widely used free radical-generating compound. Science (213): 1023-1025, 1981.
83. Ito A, Watanabe H, Naito M, Naito Y: Induction of duodenal tumors in mice by oral administration of hydrogen peroxide. Gann (72): 174-175, 1981.
84. Solanki V, Rana RS, Slaga TJ: Diminution of mouse epidermal superoxide dismutase and catalase activities by tumor promoters. Carcinogenesis (2): 1141-1146, 1981.
85. Logani MK, Solanki V, Slaga TJ: Effect of tumor promoters on lipid peroxidation in mouse skin. Carcinogenesis (3): 1303-1306, 1982.

86. Kensler TW, Bush DM, Kozumbo WJ: Inhibition of tumor promotion by a biomimetic superoxide dismutase. Science (221): 75-77, 1983.
87. Leuthauser SWC, Oberley LW, Oberley TD, Sorenson JRJ, Ramakrishna K: Antitumor effect of a copper coordination compound with superoxide dismutase-like activity. J Natl Cancer Inst (66): 1077-1081, 1981.
88. Hoffman M, Autor AP: Effect of cyclooxygenase inhibitors on phorbol-induced stimulation of oxygen consumption and superoxide production by rat pulmonary macrophages. Biochem Pharmacol (31): 775-780, 1982.
89. Kinsella AR, Gainer HST, Butler J: Investigation of a possible role for superoxide anion production in tumor promotion. Carcinogenesis (4): 717-719, 1983.
90. Carp H, Janoff A: In vitro suppression of serum elastase-inhibitor capacity by reactive oxygen species generated by phagocytosing polymorphonuclear leukocytes. J Clin Invest (63): 793-797, 1979.
91. Hopsu-Havu VK, Fraki JE, Jarvinen M: Proteolytic enzymes in the skin. In: Barrett AJ (ed) Proteinases in mammalian cells and tissues. North-Holland Publ., New York, 1979, pp 545-591.
92. Fukuyama K, Ohtani O, Hibino T, Epstein WL: Cellular localization of thiol-protease inhibitor in the epidermis of the new born rat. Cell Tissue Res (223): 313-323, 1982.
93. Shamberger RJ, Rudolph G: Increase of lysosomal enzymes in skin cancer. Nature (213): 617-618, 1967.
94. Brown RL, Clark RW, Chiu J-F, Stubblefield E: Protease activation of G1 nuclei isolated from Chinese hamster fibroblasts. Exp Cell Res (104): 207-213, 1977.
95. Harris CC, Primek A, Cohen MH: Elevated alpha$_1$-antitrypsin serum levels in lung cancer patients. Cancer Res (34): 280-281, 1974.
96. Chawla RK, Wadswork AD, Rudman D: Relation of the urinary cancer-related glycoprotein EDC1 to plasma inter-A1-trypsin inhibitor. J Immunol (121): 1636-1639, 1978.

97. Gaffar SA, Princler GL, McIntire KR, Braatz JA: A human lung tumor-associated antigen cross-reactive with α1-antichymotrypsin. J Biol Chem (255): 8334-8339, 1981.
98. Kelly UL, Cooper EH, Alexander C, Stone J: The assessment of antichymotrypsin in cancer monitoring. Biomedicine (Paris) (28): 209-215, 1978.
99. Kondo Y, Ohsawa N: Production of human α1-antichymotrypsin-like protein by a human malignant melanoma transplanted into nude mice. Cancer Res (42): 1549-1554, 1982.
100. Weissmann G, Korchak HM, Perez HD, Smolen JE, Goldstein IM, Hoffstein ST: Leukocytes as secretory organs of inflammation. In: Weissmann G, Samuelsson B, Paoletti R: (eds) Advances in inflammation research, Vol. 1, Raven Press, New York, 1978, pp 95-112.
101. Powles TJ, Bockman RS, Honn KV, Ramwell P: (eds) Prostaglandins and cancer: First Internation Conference. Alan R. Liss, New York, 1982.
102. Samuelsson B, Paoletti R: (eds) Advances in prostaglandin, thromboxane, and leukotriene research series, Vol. 9, Raven Press, New York, 1982.
103. Nakadata T, Yamamoto S, Iseki H, Sonoda S, Takemura S, Ura A, Hosoda Y, Kato R: Inhibition of 12-0-tetradecanoyl-phorbol-13-acetate-induced tumor promotion by nordihydroguaiaretic acid, a lipoxygenase inhibitor, and p-bromophenylacyl bromide, a phosholipase A_2 inhibitor. Gann (73): 841-843, 1982.
104. Levine L: Arachidonic acid transformation and tumor production. Adv Cancer Res (35): 49-79, 1982.
105. Belman S, Troll W: Hormones, cyclic nucleotides, and prostaglandins. In: Slaga TJ, Sivak A, Boutwell RK (eds) Carcinogenesis Vol. 2, Mechanisms of tumor promotion and cocarcinogenesis. Raven Press, New York, 1978, pp 117-134.
106. Chang J, Wigley F, Newcombe D: Neutral protease activation of peritoneal macrophage prostaglandin synthesis. Proc Natl Acad Sci USA (77): 4736-4740, 1980.

107. Hatcher VB, Wertheim MS, Rhee CY, Tsien G, Burk PG: Relationship between cell surface protease activity and doubling time in various normal and transformed cells. Biochim Biophys Acta (451): 499-510, 1976.
108. Hatcher VB, Oberman MS, Wertheim MS, Rhee CY, Tsien G, Burk PG: The relationship between surface activity and the rate of cell proliferation in normal and transformed cells. Biochem Biophys Res Commun (76): 602-608, 1977.
109. Markus G, Takita H, Camiolo SM, Corasanti JG, Evers JL, Hobika GH: Content and characterization of plasminogen activators in human lung tumors and normal lung tissue. Cancer Res (40): 841-848, 1980.
110. Moscattelli D, Jaffe E, Rifkin DB: Tetradecanoyl phorbol acetate stimulates latent collagenase production by cultured human endothelial cells. Cell (20): 343-351, 1980.
111. Reich E: Activation of plasminogen: A widespread mechanism for generating localized extracellular proteolysis. In: Ruddon RW (ed) Biological markers of neoplasia: Basic and applied aspects. Elsevier publ., New York, 1978, pp 491-500.
112. Long SD, Quigley JP, Troll W, Kennedy AR: Protease inhibitor antipain suppresses 12-0-tetradecanoyl-phorbol-13-acetate induction of plasminogen activator in transformable mouse embryo fibroblasts. Carcinogenesis (2): 933-936, 1981.
113. Sloane BF, Dunn JR, Honn KV: Lysosomal cathepsin B: Correlation with metastatic potential. Science (212): 1151-1153, 1981.
114. Honn KV, Cavanaugh P, Evens C, Taylor JD, Sloane BF: Tumor cell-platelet aggregation: induced by cathepsin B-like proteinase and inhibited by prostacyclin. Science (217): 540-542, 1982.
115. Hornback W, Brechmier D, Bellon G, Adnet JJ, Robert L: Biological significance of elastase-like enzymes in arteriosclerosis and human breast cancer. In: Strauli P, Barrett AJ, Baici A (eds) Proteinases and tumor invasion,

Raven Press, 1980, pp 117-141.
116. Mott DM, Fabisch PH, Sani BP, Sorof S: Lack of correlation between fibrinolysis and the transformed state of cultured mammalian cells. Biochem Biophys Res Commun (61): 621-627, 1974.
117. Plagemann PGW, Estensen RD: Lack of correlation between effects of tumor promoter TPA on plasminogen activator production, phosphatidyl choline synthesis, and hexose transport in mammalian cell culture systems. J Cell Physiol (104): 105-110, 1980.
118. Brynes PJ, Schmidt R, Hecker E: Plasminogen activator induction and platelet aggregation by phorbol and some of its derivatives: Correlation with skin irritancy and tumor-promoting activity. J Cancer Res Clin Oncol (97): 257-266, 1980.
119. Wilson EL, Jacobs P, Dowdle EB: The effects of dexamethasone and tetradecanoyl phorbol acetate on plasminogen activator release by human acute myeloid leukemia cells. Blood (61): 561-566, 1983.
120. Weinstein IB, Mufson RA, Lee L-S, Fisher PB, Laskin J, Horowitz AD, Ivanovik V: Membrane and other biochemical effects of the phorbol esters and their relevance to tumor promotion. In: Pullman B, Ts'O POP, Gelboin H (eds) Carcinogenesis: Fundamental mechanisms and environmental effects. D. Reidel Publ. Co., Boston, 1980, pp 543-563.
121. Jaken S, Black PH: Regulation of plasminogen activator in 3T3 cells: Effect of phorbol myristate acetate on subcellular distribution and molecular weight. J Cell Biol (90): 727-731, 1981.
122. Wigler M, Defeo D, Weinstein IB: Induction of plasminogen activator in cultured cells by macrocyclic plant diterpene esters and other agents related to tumor promotion. Cancer Res (38): 1434-1437, 1978.
123. Dolbeare F: Enzyme responses to tumor promoters: Cathepsin B induction in 10T½ cells. Proc Amer Assoc Cancer Res (20): 180, 1979.

124. Recklies AD, Mort JS, Poole AR: Secretion of a thiol proteinase from mouse mammary carcinomas and its characterization. Cancer Res (42): 1026-1032, 1982.
125. Steven FS, Griffin MM, Itzhaki S, Al-Habib A: A trypsin-like neutral protease on Ehrlich ascites cell surfaces: Its role in the activation of tumour-cell zymogen of collagenase. Br J Cancer (42): 712-721, 1980.
126. Tarin D, Hoyt BJ, Evans DJ: Correlation of collagenase secretion with metastatic-colonization potential in naturally occurring murine mammary tumours. Br J Cancer (46): 266-278, 1982.
127. Brinckerhoff CE, Harris Jr. ED: Modulation by retinoic acid and corticosteroids of collagenase production by rabbit synovial fibroblasts treated with phorbol myristate acetate or poly (ethylene glycol). Biochim Biophys Acta (677): 424-432, 1981.
128. Brinckerhoff CE, McMillan RM, Fahey JV, Harris Jr. ED: Collagenase production by synovial fibroblasts treated with phorbol myristate acetate. Arthritis Rheum (22): 110-116, 1979.
129. Brinckerhoff CE, Gross RH, Nagasa H, Sheldon L, Jackson RC, Harris Jr. ED: Increased level of translatable collagenase messenger ribonucleic acid in rabbit synovial fibroblasts treated with phorbol myristate acetate or crystals of monosodium urate monohydrate. Biochem (21): 2674-2679, 1982.
130. Dahlgren ME, Davies P, Bonney RJ: Phorbol myristate acetate induces the secretion of an elastase by populations of resident and elicited mouse peritoneal macrophages. Biochim Biophys Acta (630): 338-351, 1980.
131. Zucker MB, Troll W, Belman S: The tumor-promoter phorbol ester (12-0-tetradecanoyl-phorbol-13-acetate), a potent aggregating agent for blood platelets. J Cell Biol (60): 325-336, 1974.
132. Pearlstein E, Ambrogio C, Gasic G, Karpatkin S: Inhibition of the platelet-aggregating activity of two human adenocarcinomas of the colon and an anaplastic murine

tumor with a specific thrombin inhibitor, dansylarginine N-(3-ethyl-1,5-pentanediyl) amide. Cancer Res (41): 4535-4539, 1981.
133. Dabbous MKH, El-Torky M, Haney L, Brinkley Sr. B: Stimulation of collagenase release by rabbit carcinoma-derived cells. Proc Amer Assoc Cancer Res (24): 4, 1983.
134. Tarin D, Ogolvie DJ, McKinnell RG: Temperature-dependent elevation of collagenase by the renal adenocarcinoma of the leopard frog. Proc Amer Assoc Cancer Res (24): 26, 1983.
135. DiStefano JF, Beck G, Zucker S: Cancer cell membrane proteases in invasion and normal cell destruction. Proc Amer Assoc Cancer Res (24): 27, 1983.
136. Starkey JR, Hosick HL, Young DM: Comparison of basement membrane (BM) degradation and digestion of purified matrix components as correlates with metastatic/invasion behaviour of tumor cells. Proc Amer Assoc Cancer Res (24): 27, 1983.
137. Quigley JP: Morphological alterations and degradative ability of RSV-transformed chick fibroblasts when cultured in the extracellular matrix produced by normal chick fibroblasts. Proc Amer Assoc Cancer Res (24): 29, 1983.
138. Wang BS, McLoughlin GA, Richie JP, Mannick JA: Correlation of the production of plasminogen activator with tumor metastasis in B16 mouse melanoma cell lines. Cancer Res (40): 288-292, 1980.
130. Davies RL, Rifkin DB, Tepper R, Miller A, Kucherplati R: A polypeptide secreted by transformed cells that modulates human plasminogen activator function. Science (221): 171-173, 1983.
140. LeBlanc PP, Back N: Proteases during growth of Ehrlich ascites tumor. I. The fibrinolytic system. J Natl Cancer Inst (54): 881-886, 1975.
141. Glenn KC, Carney DH, Fenton II JW, Cunningham DD: Thrombin active site regions required for fibroblast receptor binding and initiation of cell division. J Biol Chem (255): 6609-6616, 1980.

142. Jaken S, Black PH: Correlation between a specific molecular weight form of plasminogen activator and metabolic activity of 3T3 cells. J Cell Biol (90): 721-726, 1981.
143. Crutchley DJ, Conanan LB, Maynard JR: Induction of plasminogen activator and prostaglandin biosynthesis in Hela cells by 12-0-tetradecanoylphorbol-13-acetate. Cancer Res (40): 849-852, 1980.
144. Wilson EL, Reich E: Plasminogen activator in chick fibroblasts: Induction of synthesis by retinoic acid; synergism with viral transformation and phorbol ester. Cell (15): 385-392, 1978.
145. Crutchley DJ, Maynard JR: Induction of plasminogen activator by 12-0-tetradecanoylphorbol-13-acetate and calcium ionophore. Biochim Biophys Acta (762): 76-85, 1983.
146. Dawamura H, Strickland JE, Yuspa SH: Inhibition of 12-0-tetradecanoylphorbol-13-acetate induction of epidermal transglutaminase activity by protease inhibitors. Cancer Res (43): 4073-4077, 1983.
147. Boynton AL, Whithead JF: The role of cyclic AMP in cell proliferation: A critical assessment of the evidence. Adv Cyc Nuc Res (15):193-195, 1983.
148. Goldberg ND, Haddox MK, Nicol SE, Glass DB, Sanford CH, Kuehl FA, Estensen, R: Biological regulation through opposing influences of cyclic GMP and cyclic AMP: The yin yang hypothesis. Adv Cyc Nuc Res (15):307-330, 1975.
149. Halprin KM: Cyclic nucleotides and epidermal cell proliferation. J Invest Derm (66):339-343, 1976.
150. Ross EM, Gilman AG: Biochemical properties of hormone sensitive adenylate cyclase. Ann Rev Biochem (49):533-564, 1980.
151. Exton JH, Harper SC: Role of cyclic AMP in the action of catecholamines on hepatic carbohydrate metabolism. Adv Cyc Nuc Res (5):519-532, 1975.

152. Samuelsson B, Granstrom E, Green K, Hamburg M, Hammarstrom S: Prostaglandin. Ann Rev Biochem (44):669-695, 1975.
153. Nimmo HG: Hormonal control of protein phosphorylation. Adv Cyc Nuc Res (8): 145-266, 1977.
154. Walsh DA: Role of the cAMP-dependent protein kinase s the transducer of cAMP action. Biochem Pharmacol (27):1801-1804, 1978.
155. Lincoln TM, Corbin JD: Characterization and biological role of the cGMP dependent protein kinase. Biochem Pharmacol (27):139-192, 1978.
156. Russell DH: Type I cyclic AMP dependent protein kinase as a positive effector of growth. Adv Cyclic Nuc Res (9):493-506, 1978.
157. Ryan WL, Heidrick ML: Role of cyclic nucleotides in cancer. Adv Cyclic Nuc Res (4):81-116, 1974.
158. Van Duuren BL: Tumor-promoting agents in two-stage carcinogenesis. Prog Exp Tumor Res (11):31-58, 1969.
159. Hecker E: Phorbol esters from croton oil, chemical nature and biological activities. Naturwissen-Schaften (54):282-284, 1967.
160. Rohrschneider LR, Boutwell RK: Phorbol esters, fatty acids and tumor promotion. Nature (243):212-213, 1973.
161. Sivak A, Ray F, Van Duuren BL: Phorbol ester tumor-promoting agents and membrane stability. Cancer Res (29):624-630, 1969.
162. O'Brien TG: The induction of ornithine decarboxylase as an early possible obligatory event in mouse skin carcinogenesis. Cancer Res (36):2644-2653, 1976.
163. O'Brien TG, Simsiman RC, Boutwell RK: Induction of the polyamine-biosynthetic enzymes in mouse epidermis by tumor-promoting agents. Cancer Res (35):2426-2433, 1975.
164. Byus CV, Russell DH: Ornithine decarboxylase activity: Control by cyclic nucleotides. Science (187):650-652, 1975.
165. Mufson RA, Astrup EG, Simsiman RC, Boutwell RK: Dissociation of increases in levels of 3'5' cyclic AMP and

3'5' cyclic GMP from induction of ornithine decarboxylase by the tumor promoter 12-O-tetradecanoyl phorbol-13 acetate in mouse epidermis in vivo. Proc Natl Acad Sci USA (74):657-661, 1977.

166. Trevillyan JM, Byus CV: Cyclic AMP and tumor promoters cause differential induction of ornithine decarboxylase and accumulation of putrescine in Chinese hamster ovary cells deficient in cyclic AMP-dependent protein kinase. Biochim Biophys Acta (762):187-197, 1983.

167. Lichti U, Gottesman HM: Genetic evidence that a phorbol ester tumor promoter stimulates ornithine decarboxylase activity by a pathway that is independent of cyclic AMP-dependent protein kinases in CHO cells. J Cell Physiol (113): 433-439, 1982.

168. Belman S, Troll W: Phorbol 12-myristate 13-acetate effect on cyclic adenosine 3',5'-monophosphate levels in mouse skin and inhibition of phorbol myristate acetate-promoted tumorigenesis by theophylline. Cancer Res (34):3446-3455, 1974.

169. Mufson RA, Simsiman RC, Boutwell RK: The effect of the phorbol ester tumor promoters on the basal and catecholamine-stimulated levels of cyclic adenosine 3',5'-monophosphate in mouse skin and epidermis in vivo. Cancer Res (37):665-669, 1977.

170. Yoshikawa K, Adachi K, Halprin KM, Levine V: Cyclic AMP in skin: Effects of acute ischeamia. Brit J Dermatol (92):249-254, 1975.

171. Solanki V, Murray AW: Decreased accumulation of cyclic adenosine 3'5'-monophosphate in "ischemic" skin after 12-O-tetradecanoyl-phorbol-13 acetate treatment. J Invest Derm (78):264-266, 1982.

172. Belman S, Troll W, Garte SJ: Effect of phorbol myristate acetate on cyclic nucleotide levels in mouse epidermis. Cancer Res (38):2978-2982, 1978.

173. Perchollet JP, Boutwell RK: Effects of 3-isobutyl-1 methylxanthine and cyclic nucleotides on 12-O-tetradecanoylphorbol-13-acetate-induced ornithine

decarboxylase activity in mouse epidermis in vivo. Cancer Res (41):3918-3926, 1981.
174. Garte SJ, Belman S: Effects of multiple phorbol myristate acetate treatments on cyclic nucleotide levels in mouse epidermis. Biochem Biophys Res Comm (84):489-494, 1978.
175. Garte SJ, Belman S: Unpublished observations.
176. Grotendorst GR, Schimmel SD: Alteration of cyclic nucleotide levels in phorbol 12-myristate 13-acetate treated myoblasts. Biochem Biophys Res Comm (93):301-307, 1980.
177. Coffey RG, Hadden JW: Phorbol myristate acetate stimulation of lymphocyte granulocyte cyclase. Biochem Biophys Res Comm (101):584-590, 1981.
178. Rochette-Egly C, Chouroulinkov I, Castagna M: Cyclic nucleotide levels in rat embryo fibroblasts treated with tumor promoting phorbol diester. J Cyclic Nuc Res (5):385-395, 1979.
179. Estensen RD, Hadden JW, Hadden EM, Touraine F, Touraine JL, Haddox MK, Goldberg ND: Phorbol myristate acetate: Effects of a tumor promoter on intracellular cyclic GMP in mouse fibroblasts and as a mitogen on human lymphocytes. In: Clarkson B, Baserga R (eds) Control of proliferation in animal cells. Cold Spring Harbor, new York, 1974, pp 627-634.
180. Grimm W, Marks F: Effect of tumor promoters on the normal and isoproterenol elevated level of adenosine 3'5'-cyclic monophosphate in mouse epidermis in vivo. Cancer Res (34):3128-3134, 1974.
181. Murray AW, Solanki V, Verma AK: Accumulation of cyclic adenosine 3'5'-monophosphate in adult and newborn mouse skin: Response to ischaemia and isoproterenol. J Invest Dermatol (68):125-127, 1977.
182. Belman S, Garte SJ: Antagonism between phorbol myristate acetate and butyric acid on isoproterenol elevation of cyclic adenosine 3',5'-monophosphate and their effects on β-adrenergic receptors in mouse epidermis.

Cancer Res (40):240-244, 1980.

183. Brostrom MA, Brostrom CA, Brotman LA, Lee CS, Wollf DF, Geller HM: Alterations of glial tumor cell Ca^{2+} metabolism and Ca^{2+}-dependent cAMP accumulation by phorbol myristate acetate. J Biol Chem (257):6758-6765, 1982.

184. Rochette-Egly C, Castagna M: A tumor promoting phorbol ester inhibits the cyclic AMP response of rat embryo fibroblasts to catecholamines and prostaglandins. Febs Lett (103):38-42, 1979.

185. Garte SJ, Belman S: Tumour promoter uncouples β-adrenergic receptor from adenyl cyclase in mouse epidermis. Nature (284):171-173, 1980.

186. Garte SJ, Belman S: Decreased β-adrenergic responsiveness in mouse epidermal papillomas during tumor promotion with phorbol myristate acetate. Cancer Lett (9):245-249, 1980.

187. Novogrodsky A, Patya M, Rubin AL, Stenzel KH: Inhibition of by phorbol myristate acetate is mediated by activated macrophages. Biochem Biophys Res Comm (104):389-393, 1982.

188. Garte SJ, Currie D, Belman S: Inhibition of β-adrenergic response in cultured epidermal cells by phorbol myristate acetate. Carcinogenesis (4):939-940, 1983.

189. Marks F, Ganss M, Grimm W: Agonist and mitogen-induced desensitization of isoproterenol-stimulated cyclic AMP formation in mouse epidermis in vivo. Biochim Biophys Acta (678):122-131, 1981.

190. Mufson A, Fischer SM, Verma AK, Gleason GL, Slaga TJ, Boutwell RK: Effects of 12-O-tetradecanoyl-13-acetate and mezerein on epidermal ornithine decarboxylase activity, isoproterenol-stimulated levels of cyclic adenosine 3':5'-monophosphate, and induction of mouse skin tumors in vivo. Cancer Res (39):4791-4795, 1979.

191. Slaga TJ, Fischer SM, Nelson K, Gleason, GL: Studies on the mechanism of skin tumor promotion: Evidence for several stages in promotion. Proc Natl Acad Sci USA

(77):3659-3663, 1980.
192. Furstenberger G, Berry DL, Sorg B, Marks F: Skin tumor promotion by phorbol esters is a two-stage process. Proc Natl Acad Sci (78):7722-7726, 1981.
193. Rochette-Egly C, Castagna M: A tumor-promoting phorbol ester inhibits the cyclic AMP response of rat embryo fibroblasts to catecholamines and prostaglandin E1. Febs Lett (103):38-42, 1979.
194. Ashenda CL, Boutwell RK: Prostaglandin E and F levels in mouse epidermis are increased by tumor-promoting phorbol esters. Biochem Biophys Res Comm (90):623-627, 1979.
195. Bresnick E, Meunier P, Lamden M: Epidermal prostaglandins after topical application of a tumor promoter. Cancer Lett (7):121-125, 1979.
196. Furstenberger G, Marks F: Early prostaglandin E synthesis is an obligatory event in the induction of cell proliferation in mouse epidermis in vivo by the phorbol ester TPA. Biochem Biophys Res Comm (2):749-756, 1980.
197. Fischer SM, Gleason GL, Hardin LG, Bohrman JS, Slaga TJ: Prostaglandin modulation of phorbol ester skin tumor promotion. Carcinogenesis (1):245-248, 1980.
198. Garte SJ, Belman S: Prostaglandins fail to elevate cyclic AMP levels in mouse epidermis in vivo and in vitro. J Invest Derm, in press.
199. Marks F, Furstenberger G, Kownatzki E: Prostaglandin E-mediated mitogenic stimulation of mouse epidermis in vivo by divalent cation ionophore A23187 and by tumor promoter 12-O-tetradecanoylphorbol-13-acetate. Cancer Res (41):696-702, 1981.
200. Kuehl FA, Cirillo VJ, Ham EA, Humes JL: the regulatory role of the prostaglandins on the cyclic 3'-5' AMP system. Adv Bioscience (9):155-172, 1972.
201. Feller N, Malachi T, Halbrecht I: Prostaglandin E_2 and cyclic AMP levels in human breast tumors. J Cancer Res Clin Oncol (93):275-280, 1979.
202. Gems D, Seitz M, Kramer W, Grimm W, Till G, Resch K:

Ionophore A23187 raises cyclic AMP levels in macrophages by stimulating prostaglandin E formation. Exper Cell Res (118):55-62, 1979.

203. Collett MS, Erikson RL: Protein kinase activity associated with the ovian sarcoma virus SRC gene product. Proc Natl Acad Sci (75):2021-2024, 1978.

204. Castagna M, Takai Y, Kaibuchi K, Sano K, Kikkawa U, Nishizuka Y: Direct activation of calcium activated phospholipid dependent protein kinase by tumor-promoting esters. J Biol Chem (257):7847-7851, 1982.

205. Niedel JE, Kuhn LJ, Vandenbark GR: Phorbol diester receptor copurifies with protein kinase C. Proc Natl Acad Sci (80):36-40, 1983.

206. Ludwig KW, Niels RM: Suppression of cyclic AMP-dependent protein kinase activity in murine melanoma cells by 12-O-tetradecanoyl-phorbol-13-acetate. Biochem Biophys Res Comm (95):296-303, 1980.

207. Murray AW, Froscio M: Effect of tumor promoters on the activity of cyclic adenosine 3':5'-monophosphate-dependent and independent protein kinases from mouse epidermis. Cancer Res (37):1360-1363, 1977.

208. Curtis GL, Stenback F, Ryan L: Inhibition of skin tumor formation with adenosine 3',5'-cyclic monophosphate in initiation -promotion carcinogenesis. Proc Am Assoc Cancer Res (15):61, 1974.

209. Perchollet JP, Boutwell RK: Effects of 3-isobutyl-1-methylxanthine and cyclic nucleotides on the biochemical processes linked to skin tumor promotion by 12-O-tetradecanoylphorbol-13-acetate. Cancer Res (41):3927-3935, 1981.

210. Simantov R, Sachs L: Enhancement of hormone action by a phorbol ester and anti-tubulin alkaloids involves different mechanisms. Biochim Biophys Acta (720):120-125, 1982.

PERSPECTIVES

T.J. Slaga and S.M. Fischer

Although the work that has been described in this volume has clearly shown the importance of arachidonate metabolism in tumor promotion and many associated cellular processes, there are still many unanswered questions. This is quite understandable since we are dealing with a very complex area of research. It is well known that tumor promotion, at least in mouse skin, is a multistage process involving many interrelated cellular processes. Since inflammation and epidermal hyperplasia are always associated with skin tumor promotion, the role of arachidonic acid metabolism in tumor promotion as well as many other cellular processes may simply be related to inflammation and hyperplasia. It should be emphasized that not all inflammatory and hyperplastic agents are skin tumor promoters (1). Current information suggests that the hyperplasia involved in skin tumor promotion may be related to a selection for a specific epidermal cell type or subpopulation that has been altered in some way by the tumor initiator (2).

The understanding of the role(s) of arachidonate metabolites in tumor promotion is complicated by the fact, as pointed out in Chapter 1, that the precise role(s) of the various arachidonate metabolites in normal physiology remain unclear. For example, what are the specific functions of the prostaglandins, thromboxanes, prostacyclins, hydroperoxy fatty acids and leukotrienes in the skin? Do any of the arachidonate metabolites promote tumors or induce inflammation and hyperplasia in the skin? As pointed out in Chapters 2 and 3, none of the prostaglandins act as skin tumor promoters or induce epidermal hyperplasia but some have an enhancing effect when given with a known tumor promoter. Because of their limited availability, the thromboxanes, prostacyclins, hydroperoxy fatty acids, and leukotrienes have not yet been investigated in this regard. The testing of these compounds would clearly aid in our understanding of their role in tumor promotion. Studies with inhibitors of the cyclooxygenase and lipoxygenase pathways have suggested that both pathways are very important in skin tumor promotion. There are some differences, however, in the relative importance of each pathway in different stocks of mice. For example in the SENCAR mouse (very sensitive to the induction of skin tumors), the lipoxygenase pathway appears to be very critical to tumor promotion since indomethacin at low doses (inhibitor of the cyclooxygenase pathway) enhances skin tumor promotion (see Chapter 2). In the CD-1 and NMRI mice, indomethacin in a dose-dependent manner inhibits skin tumor promotion (see Chapter 3). Since the SENCAR mouse was selected for increased sensitivity to two-stage skin carcinogenesis for eight generations, the increased activity of the lipoxygenase pathway (see Chapter 2) may be related to the increased sensitivity to tumor promotion. In this regard, a specific inhibitor of the lipoxygenase pathway would be very useful in determining its importance in tumor promotion.

As is quite evident throughout this volume, tumor promoters have a profound effect on cellular membranes in terms of both structure and function. Tumor promoters inhibit cell-cell communication (Chapter 7) and increase phospholipid synthesis in mouse skin as well as a number of cell culture systems (Chapter 4). The effects of phorbol ester tumor promoters on phospholipid metabolism as well as other cellular effects may stem from their interaction with a specific high affinity receptor (3). Recently, data from several laboratories suggest that this receptor is protein kinase C in which diacylglycerol is the natural ligand (Chapter 5). Figure 1 is a schematic diagram of the interaction of phorbol ester tumor promoters with this receptor and its relationship to phospholipid metabolism, calcium, and arachidonic acid metabolism. Although this receptor mechanism appears to be important in tumor promotion by phorbol esters and certain nonphorbol ester compounds (teleocidin and aplysiatoxin), there are other tumor promoters such as chrysarobin, anthralin, and benzoyl peroxide which do not interact directly with this receptor. They could, however, indirectly affect protein kinase C or possibly work down stream from the receptor. In this regard, a free radical mechanism of tumor promotion may be important either by a direct mechanism as with benyoyl peroxide or by an indirect mechanism as with phorbol esters (Chapters 2 and 6). The several chapters (2,6,8) dealing with the importance of free radicals, or more broadly the prooxidant states and tumor promotion, provide new insight into possible mechanism(s) of action of tumor promoters.

Since most of the research on the role of arachidonate metabolites in tumor promotion is derived primarily from studies using phorbol ester tumor promoters and the mouse skin model, it is very important to determine the role of these metabolites in other tumor promotion models such as liver, mammary gland, bladder, colon and esophagus. The use of other classes of promoters will clearly aid not only our understanding of the role of arachidonate metabolism in a given model of tumor promotion but also may provide an indication of the generality of its involvement.

REFERENCES

1. Slaga, T.J., Bowden, G.T. and Boutwell, R.K. (1975) Acetic acid, a potent stimulator of mouse epidermal macromolecular synthesis and hyperplasia but with weak tumor promoting ability. Journal of the National Cancer Institute 55: 983-987.

2. Slaga, T.J., Fischer, S.M., Weeks, C.E., Klein-Szanto, A.J.P. and Reiners, J. (1982) Studies on the mechanisms involved in multistage carcinogenesis in mouse skin. Journal of Cellular Biochemistry 18: 99-119.

3. Blumberg, P.M., Declos, K.B., Dumphy, W.G. and Jaken, S. (1982) Specific binding of phorbol ester tumor promoters to mouse tissues and cultured cells. In: E. Hecker (ed.) Cocarcinogenesis and Biological Effects of Tumour Promoters, pp. 519-535. Raven Press, New York.

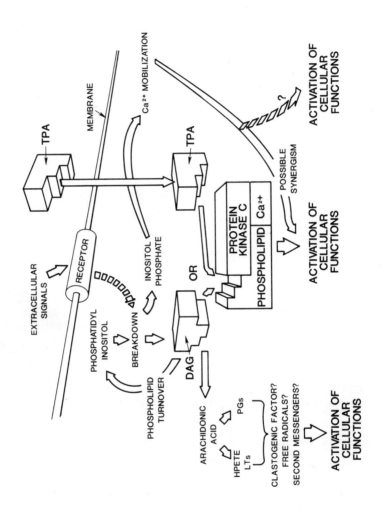

INDEX

-- A --

A23187, see calcium ionophore
Acetyl salicylic acid (aspirin), 177
Acylcarnitines, 108
Adenyl cyclase, 12, 52, 230-231
3-amino-1-n-(trifluromethyl)-phenyl-2-pyrazoline, 67
ε-aminocaproic acid (EACA), 202-203, 210-211, 220
Anthralin, 22, 218, 225
Anti-inflammatory steroids, see also dexamethasone, fluocinolone acetonide, 27, 32-33
Antioxidants, 133-135, 139-143, 147, 150, 152-154
Antipain, 207-211, 213-214, 220-227
Alkyllysophospholipid, 108
Aplysiatoxin, 103
Aprotinin, 201, 210, 212, 219, 226
Arachidonic acid, 6-14, 23-29, 34-42, 50, 55, 58, 66, 68, 74, 79-83, 85, 89-91, 93-94, 102-103, 107, 110, 117-118, 143, 148, 151-152, 174-180, 222
Ascorbate, see vitamin C

-- B --

Benoxaprofen, 38, 41
Benzilcarbonyloxy glycolate (MeDBG), 206-207
Benzoyl peroxide, 22, 33, 134, 145, 217, 219
Bowman-Birk soybean protease inhibitor (BB), 205
Bromoacetone, 204, 206
ρ-bromophenacylbromide, see dibromacetophenone
Butylated hydroxyanisole (BHA), 33, 134, 145
Butylated hydroxytoluene (BHT), 33, 134
BW755C, 24

-- C --

Calcium (Ca^{++}), 55, 91, 93, 104-119, 137, 152
Calcium ionophone (A-23187), 10, 22, 54, 63-64, 85-87, 103, 110-111, 118-119
Calmodulin, 107-108
cAMP, 12, 52, 54, 117-118, 229-232
Catalase, 40, 135-136, 139-141, 143-145, 147, 149-150, 215, 219
Cell-cell communication (see also metabolic communication) 112, 170-174, 178
cGMP, 228-229
Chalone, 119
Chick embryo skin, 13
Chloroperbenzoate, 33
Chromosomal aberrations, see also DNA damage, 132, 136 144-145, 150-151, 154, 221
Chymostatin, 207, 211, 226
Clastogenic factors (CF), 133-134, 139, 141, 146, 149-150
Copper II (3,5-diisopropylsalicylate)$_2$ (CuDIPS), 33, 37, 40, 217
Croton oil, see also phorbol esters, 134, 200, 202-203, 206
Cyclic nucleotides, see also cAMP and cGMP, 154, 227-232
Cyclooxygenase, 8-9, 12-14, 23, 25-27, 30, 68, 80, 149

-- D --

Dexamethasone, 23, 36
Diacylglyceride lipase, 118
Diacylglycerol (DAG), 82-88, 93-94, 109-114, 116-118, 141
Dibromoacetone, 209-210
Dibromoacetophenone, 24, 27-28, 30-32, 151-152
13, 14, dihydro-prostaglandins, 11
Dihydroteleocidin B, 22
5, 12-dihydroxy-eicosatetraenoic acid (5,12-DHETE), 10
N, N-dimethylamino-[ρ-(ρ´-guanidinobenzyl-oxy)], 207
DNA damage, 136-139, 144-146, 150-151, 153-154, 212, 216, 219
DNA strand breakage, 133, 215

-- E --

Ehrlich ascites cells, 210-211, 226
Ehrlich carcinoma cells, 217
Eicosatetraenoic acid, see arachidonic acid
Eicosa-5,8,11-trienoic acid, 13
5,8,11,14 eicosatetraynoic acid (ETYA), 23, 27-28, 31-32, 38, 41, 67-68, 80-81, 86, 94, 150, 177
Elastatinal, 213-214, 220

Epidermal, 15, 61, 145, 148, 152, 218, 222
Epidermal cells, 9, 23, 36, 39, 57 (see also HEL 30)
Epidermal growth factor (EGF) 57, 112, 114
Epidermis, 11, 106, 215
Epoxide hydrolase, 10
Essential fatty acid (EFA), 6-7, 11-15
Ethylphenylpropriolate, 23
-- F --
Fibroblasts, 89, 111, 132-133, 135, 217, 219, 222, 225, 229-230
   3T3 cells, 15, 57, 75, 112, 212, 214, 225
   3T6 cells, 57, 148
   PY 3T3, 221
   C3H 10T 1/2, 86, 133-135, 150, 152-153, 207-209, 224-226
Fluocinolone acetonide, 23
Fluphenazine, 109
Flurbiprofen, 26, 28, 30-32, 36, 41, 177-178
Free radicals, 33, 38-42, see also oxygen radicals, hydroxy radicals, super oxide anion, singlet oxygen
Frog bladder cells, 213
Frogs, 9
Fruit fly, 103
-- G --
Glucocorticosteroids, 118-119
α-glutamyl transpeptidase, 142
Glutathione (GSH)-peroxidase, 133-136, 139, 143, 145-147, 150, 154
Glycolipids, 111, 142
Growth factors, 103, 112, 117-118, 119, 140
Guinea pig keratinocytes, 12, 54
Guinea pigs, 9
-- H --
Hamster embryo cells, 134, 207, 209
HEL-30 cells, 55-57
HeLa cells, 75, 84-87, 175-176, 225
Hepatocytes, 150
HL-60, 91-92, 112

Human epidermis, 11, 16
Hydra, 103
Hydrogen peroxide (H_2O_2), 135 138-140, 144, 145, 150, 152, 216-217, 220
Hydroperoxides, 117, 132, 134, 136, 142, 144, 148
Hydroperoxy (acids), 6, 132, 223
Hydroperoxy-eicosatetraenoic acid (HPETE), 28-29, 33-34, 42
   5-HPETE, 143
   12-HPETE, 58
   15-HPETE, 143
Hydroxy radical (OH·), 135-137, 138, 149, 216
Hydroxy-eicosatetraenoic acid (HETE), 28, 81, 149
   12-HETE, 9, 14, 67, 141
Hydroxyfatty acids, 6, 10, 223
5-hydroxy peroxidase, 10
15-hydroxyprostaglandin dehydrogenase, 11
-- I --
Indomethacin, 10, 23-32, 35-37, 41, 51-55, 57-59, 61-63, 65-67, 79-81, 88-89, 150, 174-178, 182
Inflammation, 23, 26-27, 29, 60, 117-119
Ingenol esters, 102
-- K --
15-keto-dihydro-prostaglandin, 11
-- L --
Lauroyl peroxide, 33
Leukemia cell line, 91
Leukocytes, 39
Leukotriene B_4 (LTB_4), 11, 117
Leukotrienes, 6, 10, 12, 31, 42, 223
Leupeptin, 201-203, 207-213, 220, 223-227
Lewis lung tumors, 210-211
Linoleic acid (9,12-octadecadienoic acid), 6-7, 178-179, 182
Linolenic acid (9,12,15-octadecatrienoic acid), 6, 25-26, 178-17
Lipid peroxidation, 215-216, 218
Lipomodulin, 118
Lipoxygenase, 8-12, 14, 23-25, 149, 216, 220

Liver (see hepatocytes), 74, 138, 177, 204
Lymphocytes, 75-83, 85, 110, 112, 118, 133, 143, 145-154, 212-213, 215, 219, 229
Lymphosarcoma, 210
-- M --
Macrophages, 24, 91, 110, 143, 215-216, 219-220, 224-230
Madin Darby Canine Kidney (MDCK) cells, 89-90, 151, 175
Magnesium (Mg^{++}), 115-116
Mannitol, 40
Mellitin, 108
Mepacrine, 24-85
Metabolic communication (see also cell-cell communication), 31, 33, 179-181
4-O-methyl TPA (PMA), 54, 63-64, 92, 103, 216, 219, 226
Methyltransferase, 92-93
Mezerein, 22, 26, 36, 39, 64, 86, 102, 144, 215, 217-218, 230
Mice, 9, 13, 14, 24, 26-28
Monoacylglycerides, 111
Mouse brain, 104-105, 107
Mouse epidermal cells (keratinocytes), 10, 24, 31, 36, 39-40, 42, 58-59, 175
Mouse epidermis, 25, 29, 50, 176, 228-232
Mouse skin, 22, 29, 51, 54-56, 59-60, 63, 66-68, 102, 104, 110, 117-119, 211, 228
-- N --
NADPH-dependent PGE_2-9-keto-reductase, 11
Nematode, 103
Neutrophils, 109, 112, 118
Nordihydroguaiaretic acid (NDGA), 24-25, 27, 38, 41, 67, 81
-- O --
Octylisocyanate, 208, 210
Oleate, 110, 179
Ornithine decarboxylase (ODC), 23-27, 33, 52, 54-57, 67-68, 76, 79, 88, 112, 153-154, 175, 215, 217, 228
Oxygen radicals, 34-37, 40, 42, 132, 135-138, 140-141, 143, 148-149, 152-153, 208, 215, 223

-- P --
Palmitate, 89
Palmitoleic, 183, 185
Palmitoyl carnitine, 108
Papain, 204
Pepstatin, 210-212, 220, 226
Peptide protease inhibitor, 220
Peroxidases, 130, 135, 138, 143, 145
Peroxisomes, 139
Phenidone, 24, 27, 28, 30-32, 36, 38, 41, 67
Phenothiazine, 103, 108
Phorbol, 223
Phorbol esters, 36, 74, 109, 111-112, 114, 116-119, 172, 225

  Phorbol-13,20-diacetate, 39
  Phorbol-12,13-dibenzoate, 39
  Phorbol dibutyrate, 75, 86, 102, 115, 218
  Phorbol diester, 23
  phorbol-12-myristate-13-acetate (PMA) (see TPA)
  12-O-retinoyl phorbol-13 acetate (RPA), 60, 62-66 103, 153, 230
Phorbol ester receptor (see also protein kinase c), 102-107, 111, 115-116, 231
Phosphatidic acid, 108
Phosphatidylcholine (PC), 8, 23, 75, 76, 79, 83-94, 108
Phosphatidylethanolamine (PE), 23, 75, 89-93
Phosphatidylinositol (PI), 8, 111, 116-117
Phosphatidylserine (PS), 106, 108-110, 116
Phospholipases, 103
  Phospholipase A_2, 8, 23-24, 27, 30, 55, 66, 85, 88-89, 91, 93, 118, 141, 150-151
  Phospholipase C, 85, 87-88, 93, 108, 111-112, 141
  Phospholipase D, 87
Phospholipids, 7, 23, 24, 29, 55, 74-76, 80-83, 85, 87-88, 90-91, 94, 103, 105, 107-111, 113-114, 118
Phosphorylation, 111, 113-114
Pigs, 12

Plasminogen activator, 224-225
Platelets, 28, 108, 229
Polyamines, 108
Polymorphonuclear cells (PMNs),
  11, 33-36, 42, 133, 143, 151-152,
  154, 216, 220
Polymyxin B, 108
Polypeptides (see also protease inhibitors), 201, 205
Polyunsaturated fatty acids
  (PUFA), 6-7, 23, 109, 111, 177
Prostacyclin (PGI$_2$), 6, 8-9, 42, 80, 117
Prostaglandins (PGs), 6, 8-9,
  11-14, 23-26, 28-37, 42, 60,
  80-81, 89, 117-118, 177, 223,
  228-231

  Prostaglandin A$_1$, 80, 180, 182
  Prostaglandin A$_2$, 80
  Prostaglandin B$_1$, 80
  Prostaglandin B$_2$, 80
  Prostaglandin D$_2$ (PGD$_2$), 9, 50, 52, 55, 80
  Prostaglandin E$_1$ (PGE$_1$), 26-27, 57, 59, 80, 230-231
  Prostaglandin E$_2$ (PGE$_2$), 9,
    11-12, 14, 24-26, 29, 42,
    50-67, 80, 117, 143, 149,
    175-180, 224, 231
  Prostaglandin F$_{1\alpha}$, 80
  Prostaglandin F$_{2\alpha}$ (PGF$_{2\alpha}$),
    9, 11-12, 14-15, 24-27, 50-53,
    55-59, 61-63, 65-66, 80, 151,
    175, 177-178
  Prostaglandin G$_2$, 143, 149
Prostaglandin receptors, 14-16, 24, 52, 117, 231
Prostaglandin synthetase, 9, 150, 175
Proteases, 77, 107, 200-227
Protease inhibitors, 33, 77, 119, 132, 200-227
Protein kinase C, 105, 107-119
Psoriasis, 29
Psoriatic, 9, 66
  Plaques, 11, 14
-- Q --
Quercetin, 27, 67, 108
Quinidine, 103

-- R --
Radical scavengers, 133-134, 148
Rats, 9, 12-14
  liver, 57
  T51B liver cells, 57
  skin, 11
Retinoids (Vitamin A), 33, 40,
  60, 77-81, 89-90, 94,
  108-110, 114, 222, 225
-- S --
Sea urchin, 103
Singlet oxygen, ($^1O_2$), 136-138
Sister chromatid exchanges
  (SCE), 146, 150, 213, 219
Skin, 7-9, 13-15, 23, 53, 74,
  133-135, 138, 140, 152-154,
  200-202, 217, 221

Superoxide ($\overline{O_2}$), 112, 132-133,
  136, 139, 143, 145-149, 152,
  209, 214-220, 230
Superoxide dismutase (SOD), 33,
  37, 40, 133, 135-137, 139-142,
  144-147, 149, 151-153, 215-220
-- T --
Teleocidin, 103, 112, 114, 140,
  144, 156, 176, 181, 215-217
Tetracaine, 55, 85
12-O-tetradecanoyl phorbol-13-
  acetate (TPA), 9, 22-31, 33-36,
  38-42, 52-63, 65-68, 75-93,
  105-117, 132-153, 172-182, 200-232
Thromboxanes (TXA$_2$), 6-9, 42, 80, 117, 149, 223
Tosylarginine methyl ester (TAME),
  201-202, 220-221
Tosyllysine chloromethyl ketone (TLCK),
  201-202, 212-213, 220, 222, 226
Tosylphenylalanyl chloromethyl ketone
  (TPCK), 201-202, 212-220, 226-227
TPA promotion, 26-27, 60-61,
  117-118, 204, 206
Trasylol, 220
Triacylglyceride, 111
Trifluoperazine, 85, 87, 108, 112
Trypsin inhibitors, bovine pancreatic, 201
Trypsin inhibitors, soybean (SBTI),
  201, 204, 207-227
Tumor promoter (see also anthralin,
  benzoyl peroxide, calcium ionophore,
  mezerein, phorbol esters), 22-24,
  30-34, 42, 174, 214

Tumor promotion, 25, 28-30,
 38, 57, 61, 68, 117, 119
       -- V --
V2 carcinoma, 214
V-79 cells, 33, 146, 178-182,
 212, 215
Verapamil, 55
Vitamin C, 134-136, 139
Vitamin E, 134-136, 143-146
       -- W --
W-7 (N-(6-aminohexyl)-5-
 chloro-1-naphthenesulfonamide),
 108, 112, 179
Walker 256 carcinoma, 210